Agnes Giberne

The World's Foundations or Geology for Beginners

Second Edition

Agnes Giberne

The World's Foundations or Geology for Beginners
Second Edition

ISBN/EAN: 9783744719353

Printed in Europe, USA, Canada, Australia, Japan

Cover: Foto ©berggeist007 / pixelio.de

More available books at **www.hansebooks.com**

THE WORLD'S FOUNDATIONS

OR

GEOLOGY FOR BEGINNERS

BY

AGNES GIBERNE

AUTHOR OF 'SUN, MOON, AND STARS,' ETC.

With Sixteen Illustrations

'Of old hast Thou laid the Foundation of the Earth.'—PSA. cii. 25

SECOND EDITION

SEELEY, JACKSON, AND HALLIDAY, 54, FLEET STREET
LONDON.   MDCCCLXXXIII

All Rights Reserved.

PREFACE.

The very warm and hearty reception accorded to my little book on Astronomy, has been my best encouragement in entering upon the domain of the sister science, Geology.

This companion-volume to 'Sun, Moon, and Stars' is written upon much the same plan, and is intended for the same class of readers—for Beginners of all kinds, whether poor or rich, whether boys, girls, or grown-up people.

My object in writing it has been not so much to supply a certain amount of technical knowledge,—for this may be easily obtained from ordinary class-books,—as to open the eyes of others to the hidden wonders and possibilities of enjoyment which lie folded in this little-studied branch of science.

Geology is counted by many to be a dull subject. But if it has its dry bones, it has also its forms of poetic beauty, its scenes of loveliness, its chords of sublime harmony.

Geology is counted by others to be a dangerous subject. But if so, the danger lies in ourselves, not in Geology. Man's haste in decision, and his readiness to put faith in unproved theories, may lead him astray. The study of God's truths, if rightly undertaken, cannot cause his feet to wander.

Geology speaks to us, as surely as the Bible itself speaks to us, of the Creator and His ways, albeit in terms more ambiguous, in language more easily misunderstood. The one is His Word, the other is His Handiwork. That the one should contradict the other is not possible. That the one and the other should contain mysteries past our power to fathom, is only what we might expect from the Word and the Handiwork of an Infinite God.

I have merely to state, in conclusion, that neither time nor pains have been spared in the endeavour to ensure accuracy as well as interest. The leading Geological writers of England and of America have been my authorities. Thanks, lastly, are due for the kind and able criticisms of competent friends, who have generously given time and thought to the examination of my proof-sheets.

WORTON HOUSE, EASTBOURNE,
August, 1881.

CONTENTS.

PART I.
HOW TO READ THE RECORD.

CHAPTER	PAGE
I. WHAT THE EARTH'S CRUST IS MADE OF	1
II. WATER-BUILT ROCKS	10
III. FOSSILS IN THE ROCKS	20
IV. FIRE-BUILT ROCKS	32
V. WHAT ROCKS ARE MADE OF	42
VI. ROCK-LAYERS	53
VII. ROCK-BENDINGS	61
VIII. ICE-WORK	70
IX. THE TWO BOOKS	83

PART II.
A STORY OF OLDEN DAYS.

X. TWO KINGDOMS	91
XI. EARLIEST AGES	101
XII. THE AGE OF LOWER ANIMALS	107
XIII. THE AGE OF FISHES	118
XIV. THE AGE OF COAL	125
XV. MORE ABOUT THE AGE OF COAL	135
XVI. THE AGE OF REPTILES	143

CHAPTER		PAGE
XVII. THE AGE OF CHALK		154
XVIII. THE AGE OF MAMMALS		169
XIX. MORE ABOUT THE AGE OF MAMMALS		175
XX. THE AGE OF ICE		183
XXI. THE AGE OF MAN		195
XXII. THE TWO RECORDS		203

PART III.
THE PAST IN THE LIGHT OF THE PRESENT.

XXIII. RIVERS		219
XXIV. WATERS		228
XXV. DELTAS		236
XXVI. GLACIERS		246
XXVII. VOLCANOES		255
XXVIII. EARTHQUAKES		270
XXIX. HOT SPRINGS		281
XXX. CORAL		291
XXXI. STALACTITE		305

TABLE OF SUBJECTS.

PART I.

CHAP. I.—What the Earth is made of.—Meaning of Geology.—The Earth's Crust.—What is meant by a 'theory.'—Different kinds of Rock.—The Volume of Geology   3—9

CHAP. II.—Stratified and Unstratified Rocks.—Crystallization. What is meant by Stratified Rocks.—Changes in the Earth-crust.—Crumbling away of cliffs.—Work of torrents.—Sediment and detritus.—Sedimentary Rocks - 10—19

CHAP. III.—Building up of Stratified Rocks.—Chalk cliffs and old sea-beaches.—Land-sinkings.—Past ages of preparation.—Aqueous Rocks.—Fossiliferous Rocks.—Different kinds of fossils.—Fossils on mountain-tops, and how they came there.—Arrangement of Strata.—Disturbances in the Earth-crust - - - , - - - - - 20—31

CHAP. IV.—Volcanic Rocks, Plutonic Rocks, and Metamorphic Rocks.—Aqueous and Igneous Rocks.—Primary, Secondary, and Transition Rocks.—Building of the Crust.—Powers of Nature.—Under-ground heat.—Volcanic eruptions.—Earthquakes.—Theories of explanation, Fire and Water - - - - - - - - - 32—41

Table of Subjects.

CHAP. V.—Flint, Clay, and Lime Rocks.—What Rocks are made of.—Rhizopods.—Chalk Cliffs.—Limestone.—Fossils in Rocks.—Diatoms.—Mountain Meal.—Coal - 42—52

CHAP. VI.—Rock-building in past ages.—Arrangement of Strata.—Different kinds of fossils in different Strata.—Classification of Strata.—TABLE OF STRATA - 53—60

CHAP. VII.—Thickness of Stratified Rocks.—Chain of records.—Difficulty of reading the record.—'Stratum' and 'layer.'—'Formation.'—Rock-foldings and bendings.—'Faults' 61—69

CHAP. VIII.—Fossil rain-prints and ripple-marks.—Stones scratched and scored.—Scotch 'till.'—Boulder-clay or drift.—Erratics.—Diluvial Soil.—Glaciers.—Moraines.—Icebergs.—Probable explanation of scratched stones.—Glacial Age - - - - - - - 70—82

CHAP. IX.—World-history before the time of Adam.—Past Ages.—Days of Creation.—The two great Books.—How to read them.—Preparation of the Earth for Man 83—88

PART II.

CHAP. X.—The Animal and Vegetable Kingdoms.—Divisions in Nature.—Classification.—Animals, plants, and rocks.—Life.—'Organs.'—Sub-kingdoms and classes.—TABLE OF THE ANIMAL KINGDOM.—TABLE OF THE VEGETABLE KINGDOM - - - - - - - 91—100

CHAP. XI.—Creation of the world.—Stages in existence of heavenly bodies.—Probable stages in preparation of the earth.—Earliest-known rocks.—Earth in those days
101—106

CHAP. XII.—Primary or 'Ancient Animal' Rocks.—Three great Ages.—Age of Lower Animals or of Limestone-building.—Seas and continents.—Coral-making.—Disturbances.—Sea-weeds and sea-creatures of Silurian Days.—Trilobites.—The first Fishes - - - - - 107—117

CHAP. XIII.—Age of Fishes.—New Red Sandstone.—Coral.—Flowerless plants.—Forests.—Insects.—Fish-fossils.—Trilobites and lobsters - - - - - 118—124

CHAP. XIV.—Day of Coal-preparation.—Description of forest-scene in Coal-Age.—First reptiles.—Rising and sinking of the ground.—Coal-seams.—Sigillaria and Stigmaria.—Tree-trunks in mines - - - - - - 125—134

CHAP. XV.—Coal-beds in South Wales.—Alternate layers of rock.—Climate in Coal-Age.—Forests.—Ferns.—Trees.—Vegetable fossils in coal.—Sigillaria.—Amphibians.—First True Reptiles.—Mountains upheaved.—End of Primary Period - - - - - - - - 135—142

CHAP. XVI.—Middle-Life Period begun.—Reptiles.—Gigantic crocodiles, lizards, winged reptiles, water reptiles, etc.—Saurians.—Tracks of Amphibians and Reptiles.—Coal-preparation.—Coral.—Fishes.—Ammonites.—Bird-skeleton.—First Mammals.—Plants - - - - 143—153

CHAP. XVII.—Chalk-building.—Great Chalk formation.—Reptiles.—Animalcules.—Nature of mud at bottom of Atlantic Ocean.—Rhizopods and Diatoms.—Chalk and flint.—Time occupied in building.—Chalk-formation in America.—Geography of the Age.—Great change in plants.—Flowering plants.—Reptiles, birds, and ammonites.—Close of Chalk-Age and Secondary Period.—Widespread destruction of life.—Possible Ice-action - - - - 154—168

CHAP. XVIII.—New-Life Period begun.—Sea and freshwater shells.—Third-Period rocks and fossils.—Nummulites.—Upheaval of mountains.—Climate.—Plants.—London and Paris Basins.—British Isles - - - 169—174

CHAP. XIX.—Possible Icebergs.—Signs of cold.—Mammals.—Ammonites, nummulites and corals.—Vegetation.—Animals in England and France.—Ancient quadrupeds and whales.—European plants and trees - - - - 175—182

CHAP. XX.—Post-Tertiary Age.—Great Ice-Age.—Drift or Till.—Periods and Ages.—Glacier Theory.—Huge quadrupeds.—Time of Floods.—Mammoths and mastodons 183—194

CHAP. XXI.—Human relics.—Stone-Age, Bronze-Age, and Iron-Age.—History of Nations.—Human remains in caves.—Supposed second Ice-Age - - - - 195—202

CHAP. XXII.—The Divine Architect.—The Bible-Record of Creation.—Theories of explanation.—Water and Fire.—A caution.—The uniformity theory.—Periods and Ages.—
TABLE OF GEOLOGICAL AGES - - - 203—216

PART III.

CHAP. XXIII.—Work of running water.—Wearing away of rock.—Earl of Mar's Punch-bowl.—Pot-holes.—'Glacier-holes' at Lucerne.—Reuss in Pass of St. Gothard.—Landslip on the Vispbach.—River Simeto.—Niagara.—Two beds of a river - - - - - - 219—227

CHAP. XXIV.—Rains.—Tropical rainfalls.—Artesian wells.—Wear of cliffs round Britain.—Shetland Islands.—Bell-Rock lighthouse.—Norfolk.—Sussex - - - 228—235

CHAP. XXV.—Sand-bars.—Mud-banks.—River Deltas—Lake and Ocean Deltas—Lake of Geneva.—The Rhone.—The Po and Adige.—The Nile.—The Ganges and Brahmapootra.—Mississippi.—Bay of Fundy mud-flats - - 236—245

Chap. XXVI.—A glacier.—Movements of a glacier.—Icebergs.
—Mont Blanc glaciers.—The Mer de Glace.—Greenland
glaciers.—Moraines.—Spitzbergen icebergs.—Greenland
ice-foot - - - - - - - 246—254

Chap. XXVII.—A Volcano.—A crater.—A cone.—Active,
Dormant, and Extinct Volcanoes.—Different kinds of eruptions.—Volcanoes of the Andes.—Mexican volcanoes.—
Icelandic volcano.—Etna.—Galongoon.—Vesuvius.—
Hawaiian volcanoes.—Monte Nuovo.—Lakes of fire.—
Fountains of fire - - - - - - 255—269

Chap. XXVIII.—Cause of Earthquakes.—New Zealand earthquakes.—A fault.—Earthquake in Chili.—Earthquake in
Cutch.—Caraccas earthquake.—Calabrian earthquake.—
Waves and cracks.—Earthquake of Lisbon.—Earthquake
in Jamaica.—General effects of earthquakes - 270—280

Chap. XXIX.—Slow movements of earth-crust.—Rising of land
in Sweden.—Sinking of land in Greenland.—Floor of
Pacific Ocean supposed to sink.—Temple of Jupiter Serapis
—Nearness of volcanoes to ocean.—Power of steam.—Hot
springs.—Bath.—Geysers.—Icelandic geysers.—American
geysers - - - - - - - 281—290

Chap. XXX.—The coral-Polyp.—Kinds of Coral.—Deep-sea
corals.—Reef-building corals.—Work done by waves.—
Formation of an Island.—Atolls.—Lagoons.—Maldive
Islands.—Fringing-reefs.—Barrier-reefs.—Feejee Islands.
Theory of explanation.—Sinking sea-bottom.—Speed of
coral-formation - - - - - - 291—304

Chap. XXXI.—Peat-mosses.—Speed of Peat-formation.—Landslips.—Slide on the Rossberg.—Stalactite and Stalagmite
caves.—Digging out of passages by water.—Cheddar
Caverns.—Animal remains in caves.—How long there?—
Speed of Stalagmite formation.—Need for caution.—Much
that cannot yet be understood in Geology - 305—313

LIST OF ILLUSTRATIONS.

MER DE GLACE	*Frontispiece*
CHART OF STRATA	62
MAGNIFIED SPECK OF EARTH, AND SCRATCHED STONES	74
GLACIER	78
FOREST OF THE CARBONIFEROUS AGE	128
FERN FOSSILS IN COAL	138
FOSSILS	170
SKELETON OF MEGATHERIUM	188
ANIMALS OF THE TERTIARY AGE	192
FLOATING ICEBERG	248
GREENLAND GLACIERS	252
VOLCANO	258
EARTHQUAKE AT LISBON	276
GEYSERS	288
CORAL ISLAND	296
STALACTITE CAVERN	308

PART I.

HOW TO READ THE RECORD.

CHAPTER I.

WHAT THE EARTH'S CRUST IS MADE OF.

'Stand still and consider the wondrous works of God.'—JOB xxxvii. 14.

WHAT is the earth made of—this round earth upon which we human beings live and move?

A question more easily asked than answered, as regards a very large portion of it. For the earth is a huge ball nearly eight thousand miles in diameter, and we who dwell on the outside have no means of getting down more than a very little way below the surface. So it is quite impossible for us to speak positively as to the inside of the earth, and what it is made of. Some people believe the earth's inside to be hard and solid, while others believe it to be one enormous lake or furnace of fiery melted rock. But nobody really knows.

4 What the Earth's Crust is Made of.

If we break up the word GEOLOGY, we find that it is made out of two Greek words, *ge*, the earth, and *logos*, a word or discourse. The meaning of 'geology' is simply a word, or a discourse, or teaching, about the earth. More strictly, it is not teaching about the whole earth, but only about the crust of the earth.

This outside crust has been reckoned to be of many different thicknesses. One man will say it is ten miles thick, and another will rate it at four hundred miles. So far as regards man's knowledge of it, gained from mining, from boring, from examination of rocks, and from reasoning out all that may be learnt by these observations, we shall allow an ample margin if we count the field of geology to extend some twenty miles downwards from the highest mountain-tops. Beyond this we find ourselves in a land of darkness and conjecture.

Twenty miles is only one four-hundredth part of the earth's diameter—a mere thin shell over a massive globe. If the earth were brought down in size to an ordinary large school globe, a piece of rough brown paper covering it might well represent the thickness of this earth-crust, with which the science of geology has to do. And the whole of the globe, this earth of ours, is but one tiny planet in the great Solar System. And the centre of that Solar System, the blazing sun, though equal in size to more than a million earths, is yet himself but one star amid millions of twinkling stars, scattered broadcast through the universe. So it would seem at first sight that the field of geology is a small field compared with that of astronomy.

But as we go on we shall find that the lesser things of God are, in their way, as great as those things which at first sight may seem the mightier. We shall find prospects of wonder and power, of mystery and beauty, unfolding before our eyes. Wrapped up in the crust of this earth are marvellous tokens of the goodness and greatness of God, and strange histories of olden days are written in her stones.

So it is distinctly with the question of the earth's crust that we are now concerned—how that crust was formed, how it was changed and modified by various influences, how it was gradually built up into its present form.

For the earth's crust was not always such as it is now.

'In the beginning God made the earth'—but He did not make it then and at once complete. There was a time when man's foot had never pressed her soil. There was a time, farther back, when no wild beasts roved through her forests and no cattle browsed upon her hills. There was a time, yet farther back, when no fishes swam in her wide waters. There was a time, still farther back, when no forests clothed her mountains or lined her valleys. There was a time, yet farther removed from the present, when no mountains towered heavenward, and no valleys had been scooped out between them. There was a time, still more distant, when no continents or islands had risen out of the mighty and dark ocean, and in all the dreary waste of waters *life* was a thing unknown.

For stage by stage, God was slowly preparing this earth-crust to be the abode of man, working calmly and deliberately, as God does work, with none of the feverish haste and restless impatience of man. What are countless ages in their flight to Him who is the King of Eternity?

But now we come back to the question with which we began, the question as to what this earth is made of?

With regard to the great bulk of the globe little can be said. Very probably it is formed through and through of the same materials as the crust. This we do not know. Neither can we tell, even if it be so formed, whether the said materials are solid and cold like the outside crust, or whether they are liquid with heat. The belief has been long and widely held that the whole inside of the earth is one vast lake or furnace of melted fiery-hot material, with only a thin cooled crust covering it. Some in the present day are inclined to question this, and hold rather that the earth is solid and cold throughout, though with large lakes of liquid fire here and there, under or in her crust, from which our volcanoes are fed. Either opinion or both opinions may be mistaken.

It will be found, as we go on, that a great many opposite opinions are held on many questions in geology, and that a great many theories are started which have very little real foundation. A 'theory' is a *possible* explanation of a mystery, put forward as the best which can be offered, until something more shall be known about the matter. Some theories are in time found to be the true explanation, but a great many more have to be thrown aside. No theory should ever be spoken of as if it were a TRUTH, until it is plainly proved to be such, with no possibility of mistake.

So with regard to the earth's inside and what it is made of, we cannot get beyond theories. But when we speak of the earth's crust, the real domain of geology, there is not the same sort of difficulty, since here we can see, and feel, and examine for ourselves.

The materials of which the crust is made are many and various, yet, generally speaking, they may all be classed under one simple word, and that word is— ROCK.

It must be understood that, when we talk of rock in this geological sense, we do not only mean hard and solid stone, as in common conversation. Rock may be changed by heat into a liquid or 'molten' state, as ice is changed by heat to water. Liquid rock may be changed by yet greater heat to vapour, as water is changed to steam, only we have in a common way no such heat at command as would be needed to effect this. Rock may be hard or soft. Rock may be chalky, clayey, or sandy. Rock may be so close-grained that strong force is needed to break it; or it may be so porous—so full of tiny holes—that water will drain through it; or it may be crushed and crumbled into loose grains, among which you can pass your fingers.

The cliffs above our beaches are rock; the sand upon our seashore is rock; the clay used in brick-making is rock; the limestone of the quarry is rock; the marble of which our mantel-pieces are made is rock. The soft sandstone of South Devon, and the hard granite of the north of Scotland, are alike rock. The pebbles in the road are rock; the very mould in our gardens is largely composed of crumbled rock.

So the word in its geological sense is a word of wide meaning.

Now the business of the geologist is to read the history of the past in these rocks of which the earth's crust is made. This may seem a singular thing to do, and I can assure you it is not an easy task.

For, to begin with, the history itself is written in a strange language, a language which man is only just beginning to spell out and understand. And this is only half the difficulty with which we have to struggle.

If a large and learned book were put before you, and you were set to read it through, you would, perhaps, have no insurmountable difficulty, with patience and perseverance, in mastering its meaning.

But how if the book were first chopped up into pieces, if part of it were flung away out of reach, if part of it were crushed into a pulp, if the numbering of the pages were in many places lost, if the whole were mixed up in confusion, and if *then* you were desired to sort, and arrange, and study the volume?

Picture to yourself what sort of a task this would be, and you will have some idea of the labours of the patient geologist.

CHAPTER II.

WATER-BUILT ROCKS.

'The waters wear the stones.'—JOB xiv. 19.

ROCKS may be divided into several kinds or classes. For the present moment it will be enough to consider the two grand divisions—STRATIFIED ROCKS and UNSTRATIFIED ROCKS.

Unstratified rocks are those which were once, at a time more or less distant, in a melted state from intense heat, and which have since cooled into a half *crystallized* state; much the same as water, when growing colder, cools and crystallizes into ice. Strictly speaking, ice is rock, just as much as granite and sandstone are rock. Water itself is of the nature of rock, only as we commonly know it in the liquid state we do not commonly call it so.

'Crystallization' means those particular forms or

shapes in which the particles of a liquid arrange themselves, as that liquid hardens into a solid—in other words, as it freezes. Granite, iron, marble, are frozen substances, just as truly as ice is a frozen substance; for with greater heat they would all become liquid like water. When a liquid freezes, there are always crystals formed, though these are not always visible without the help of a microscope. Also the crystals are of different shapes with different substances.

If you examine the surface of a puddle or pond, when a thin covering of ice is beginning to form, you will be able to see plainly the delicate sharp needle-like forms of the ice crystals. Break a piece of ice, and you will find that it will not easily break just in any way that you may choose, but will only split along the lines of these needle-like crystals. This particular mode of splitting in a crystallized rock is called the *cleavage* of that rock.

Crystallization may take place either slowly or quickly, and either in the open air or far below ground. The lava from a volcano is an example of rock which has crystallized in the open air; and granite is an example of rock which has crystallized underground beneath great pressure.

Stratified rocks, on the contrary, which make up a

very large part of the earth's crust, are not crystallized. Instead of having cooled from a liquid into a solid state, they have been slowly *built up*, bit by bit and grain upon grain, into their present form, through long ages of the world's history. The materials of which they are made were probably once, long long ago, the crumblings from granite and other crystallized rocks, but they show now no signs of crystallization.

They are called 'stratified' because they are in themselves made up of distinct layers, and also because they lie thus one upon another in layers, or *strata*, just as the leaves of a book lie, or as the bricks of a house are placed.

Throughout the greater part of Europe, of Asia, of Africa, of North and South America, of Australia, these rocks are to be found, stretching over hundreds of miles together, north, south, east, and west, extending up to the tops of some of the earth's highest mountains, reaching down deep into the earth's crust. In many parts, if you could dig straight downwards through the earth for thousands of feet, you would come to layer after layer of these stratified rocks, one kind below another, some layers thick, some layers thin, here a stratum of gravel, there a stratum of sandstone, here a stratum of coal, there a stratum of clay.

But how, when, where, did the building up of all these rock-layers take place?

People are rather apt to think of land and water on the earth as if they were fixed in one changeless form,—as if every continent and every island were of exactly the same shape and size now that it always has been and always will be.

Yet nothing can be further from the truth. The earth-crust is a scene of perpetual change, of perpetual struggle, of perpetual building up, of perpetual wearing away.

The work may go on slowly, but it does go on. The sea is always fighting against the land, beating down her cliffs, eating into her shores, swallowing bit by bit of solid earth; and rain and frost and inland streams are always busily at work, helping the ocean in her work of destruction. Year by year and century by century it continues. Not a country in the world which is bordered by the open sea has precisely the same coast-line that it had one hundred years ago; not a land in the world but parts each century with masses of its material, washed piecemeal away into the ocean.

Is this hard to believe? Look at the crumbling cliffs around old England's shores. See the effect upon the beach of one night's fierce storm. Mark the pathway on the cliff, how it seems to have crept so near the edge that here and there it is scarcely safe to tread ; and very soon, as we know, it will become impassable. Just from a mere accident, of course,— the breaking away of some of the earth, loosened by rain and frost and wind. But this is an accident which happens daily in hundreds of places around our shores.

Leaving the ocean, look now at this river in our neighbourhood, and see the slight muddiness which seems to colour its waters. What from ? Only a little earth and sand carried off from the banks as it flowed—very unimportant and small in quantity, doubtless, just at this moment and just at this spot. But what of that little going on week after week, and century after century, throughout the whole course of the river ; ay, and throughout the whole course of every river and rivulet in our whole country and in every other country. A vast amount of material must every year be thus torn from the land and given to the ocean. For the land's loss here is the ocean's gain.

And, strange to say, we shall find that this same ocean, so busily engaged with the help of its tributary rivers in pulling down land, is no less busily engaged with their help in building it up.

You have sometimes seen directions upon a phial of medicine to 'shake' before taking the dose. When you have so shaken the bottle the clear liquid grows thick; and if you let it stand for a while the thickness goes off, and a fine grain-like or dust-like substance settles down at the bottom—the settlement or *sediment* of the medicine. The finer this sediment, the slower it is in settling. If you were to keep the liquid in gentle motion, the fine sediment would not settle down at the bottom. With coarser and heavier grains the motion would have to be quicker to keep them supported in the water.

Now it is just the same thing with our rivers and streams. Running water can support and carry along sand and earth, which in still water would quickly sink to the bottom; and the more rapid the movement of the water, the greater is the weight it is able to bear.

This is plainly to be seen in the case of a mountain torrent. As it foams fiercely through its rocky bed it bears along, not only mud and sand and gravel, but stones and even small rocks, grinding the latter roughly together till they are gradually worn away, first to rounded pebbles, then to sand, and finally to mud. The material thus swept away by a stream, ground fine, and carried out to sea—part being dropped by the way on the river-bed—is called *detritus*, which simply means *worn-out* material.

The tremendous carrying-power of a mountain torrent can scarcely be realized by those who have not observed it for themselves. I have seen a little mountain-stream swell in the course of a heavy thunderstorm to such a torrent, brown and turbid with earth torn from the mountain-side, and sweeping resistlessly along in its career a shower of stones and rock-fragments. That which happens thus occasionally with many streams is more or less the work all the year round of many more.

As the torrent grows less rapid, lower down in its course, it ceases to carry rocks and stones, though the grinding and wearing away of stones upon the rocky bed continues, and coarse gravel is borne still upon its waters. Presently the widening stream, flowing yet more calmly, drops upon its bed all such coarser gravel as is not worn away to fine earth, but still bears on the lighter grains of sand. Next the slackening speed makes even the sand too heavy a weight, and that in turn falls to line the river-bed, while the now broad and placid stream carries only the finer particles of mud suspended in its waters. Soon it reaches the ocean, and the flow being there checked by the incoming ocean-tide, even the mud can no longer be held up, and it also sinks slowly in the shallows near the shore, forming sometimes broad mud-banks dangerous to the mariner.

This is the case only with smaller rivers. Where the stream is stronger, the mud-banks are often formed much farther out at sea; and more often still the river-detritus is carried away and shed over the ocean-bed, beyond reach of our ken. The powerful rush of water in earth's greater streams bears enormous masses of sand and mud each year far out into the ocean, there dropping quietly the gravel, sand, and earth, layer upon layer at the bottom of the sea. Thus pulling down and building up go on ever side by side; and while land is the theatre oftentimes of decay and loss, ocean is the theatre oftentimes of renewal and gain.

Did you notice the word *Sediment* used a few pages back about the settlement at the bottom of a medicine-phial?

There is a second name given to the Stratified Rocks, of which the earth's crust is so largely made up. They are called also SEDIMENTARY ROCKS.

The reason is simply this. The Stratified Rocks of the present day were once upon a time made up out of the *sediment* stolen first from land and then allowed to settle down on the sea-bottom.

Long, long ago the rivers, the streams, the ocean, were at work, as they are now, carrying away rock and gravel, sand and earth. Then, as now, all this material, borne upon the rivers, washed to and fro by the ocean, settled down at the mouths of rivers or at the bottom of the sea, into a *sediment*, one layer forming over another, gradually built up through long ages. At first it was only a soft loose sandy or muddy sediment, such as you may see on the sea-shore, or in a mud-bank. But as the thickness of the sediment increased, the weight of the layers above gradually pressed the lower layers into firm hard rocks; and still, as the work of building went on, these layers were, in their turn, made solid by the increasing weight over them. Certain chemical changes had also a share in the transformation from soft mud to hard rock, which need not be here considered.

All this has through thousands of years been going on. The land is perpetually crumbling away; and fresh land under the sea is being perpetually built up, from the very same materials which the sea and the rivers have so mercilessly stolen from continents and islands. This is the way, if geologists rightly judge, in which a very large part of the enormous formations of Stratified or Sedimentary Rocks have been made.

So far is clear. But now we come to a difficulty.

CHAPTER III.

FOSSILS IN THE ROCKS.

'Dead things are formed from under the waters with the inhabitants thereof.'—Job xxvi. 5 (marg.).

THE Stratified Rocks, of which a very large part of the continents is made, appear to have been built up slowly, layer upon layer, out of the gravel, sand, and mud, washed away from the land and dropped on the floor of the ocean.

You may see these layers for yourself as you walk out into the country. Look at the first piece of bluff rock you come near, and observe the clear pencil-like markings of layer above layer—not often indeed lying *flat*, one over another, and this must be explained later, but however irregularly slanting, still plainly visible. You can examine these lines of stratification on the nearest cliff, the nearest quarry, the nearest bare headland, in your neighbourhood.

But how can this be? If all these stratified rocks are built on the floor of the ocean out of material taken *from* the land, how can we by any possibility find such rocks *upon* the land? In the beds of rivers we might indeed expect to see them, but surely nowhere else save under ocean waters.

Yet find them we do. Through England, through the two great world-continents, they abound on every side. Thousands of miles in unbroken succession are composed of such rocks.

Stand with me near the sea-shore, and let us look around. Those white chalk cliffs—they, at least, are not formed of sand or earth. True, and the lines of stratification are in them very indistinct, if seen at all; yet they too are built up of sediment of a different kind, dropping upon ocean's floor. More of this later. See, however, in the rough sides of yonder bluff the markings spoken of, fine lines running alongside of one another, sometimes flat, sometimes bent or slanting, but always giving the impression of layer piled upon layer. Yet how can one for a moment suppose that the ocean-waters ever rose so high?

Stay a moment. Look again at yonder white chalk cliff, and observe a little way below the top a singular band of shingles, squeezed into the cliff, as it were, with chalk below and earth above.

That is believed to be an old sea-beach. Once upon a time the waters of the sea are supposed to have washed those shingles, as now they wash the shore near which we stand, and all the white cliff must have lain then beneath the ocean.

Geologists were for a long while sorely puzzled to account for these old sea-beaches, found high up in the cliffs around our land in many different places.

They had at first a theory that the sea must once, in far-back ages, have been a great deal higher than it is now. But this explanation only brought about fresh difficulties. It is quite impossible that the level of the sea should be higher in one part of the world than in another. If the sea around England were then one or two hundred feet higher than it is now, it must have been one or two hundred feet higher in every part of the world where the ocean-waters have free flow. One is rather puzzled to know where all the water could have come from, for such a tremendous additional amount. Besides, in some places remains of sea-animals are found in mountain heights, as much as two or three thousand feet above the sea-level—as, for instance, in Corsica. This very much increases the difficulty of the above explanation.

So another theory was started instead, and this is now generally supposed to be the true one. What if, instead of the whole ocean having been higher, parts of the land were lower? England at one time, parts of Europe at another time, parts of Asia and America at other times, may have slowly sunk beneath the ocean, and after long remaining there have slowly risen again.

This is by no means so wild a supposition as it may seem when first heard, and as it doubtless did seem when first proposed. For even in the present day these movements of the solid crust of our earth are going on. The coasts of Sweden and Finland have long been slowly and steadily rising out of the sea, so that the waves can no longer reach so high upon those shores as in years gone by they used to reach. In Greenland, on the contrary, land has long been slowly and steadily sinking, so that what used to be the shore now lies under the sea. Other such risings and sinkings might be mentioned, as also many more in connection with volcanoes and earthquakes, which are neither slow nor steady, but sudden and violent.

So it becomes no impossible matter to believe that, in the course of ages past, all those wide reaches of our continents and islands, where sedimentary rocks are to be found, were each in turn, at one time or another, during long periods, beneath the rolling waters of the ocean.

It should be clearly understood that, in speaking of those probable long-past ages of slow preparation, about which the rocks seem to tell us, I am not in anywise touching upon the question of the flood upon the earth in the time of Noah. That was comparatively a recent event. Whether it was entirely miraculous, or whether it was in part caused by some great and rapid sinking of the land, is an interesting consideration, with which, however, we are at this moment not concerned. The slow building up of the continents under the ocean could have taken place in no such brief space of time as the few months of the deluge in Noah's days, but must have occupied very long periods.

These built-up rocks are not only called 'Stratified,' and 'Sedimentary.' They have also the name of 'AQUEOUS ROCKS,' from the Latin word *aqua, water;* because they are believed to have been formed by the action of water.

They have yet another and fourth title, which is, 'FOSSILIFEROUS ROCKS.'

Fossils are the hardened remains of animals and vegetables found in rocks. They are rarely, if ever, seen in unstratified rocks; but many layers of stratified rocks abound in these remains. Whole skeletons as well as single bones, whole tree-trunks as well as single leaves, are found thus embedded in rock-layers, where in ages past the animal or plant died and found a grave. They exist by thousands in many parts of the world, varying in size from the huge skeleton of the elephant to the tiny shell of the microscopic animalcule.

Fossils differ greatly in kind. Sometimes the entire shell or bone is changed into stone, losing all its animal substance, but retaining its old outline and its natural markings. Sometimes the fossil is merely the hardened impress of the outside of a shell or leaf, which has dented its picture on soft clay, and has itself disappeared, while the soft clay has become rock, and the indented picture remains fixed through after-centuries. Sometimes the fossil is the cast of the inside of a shell; the said shell having been filled with soft mud, which has taken its exact shape and hardened, while the shell itself has vanished. The most complete description of fossil is the first of these three kinds. It is wonderfully shown sometimes in fossil wood, where all the tiny cells and delicate fibres remain distinctly marked as of old, only the whole woody substance has changed into hard stone.

But although the fossil remains of quadrupeds

and other land-animals are found in large quantities, still their number is small compared with the enormous amount of fossil sea-shells and sea-animals.

Land-animals can, as a rule, have been so preserved, only when they have been drowned in ponds or rivers, or mired in bogs and swamps, or overtaken by frost, or swept out to sea.

Sea-animals, on the contrary, have been so preserved on land whenever that land has been under the sea ; and this appears to have been the case, at one or another past age, with the greater part of our present continents. These fossil remains of sea-animals are discovered in all quarters of the world, not only on the sea-shore but also far inland, not only deep down underground but also high up on the tops of lofty mountains—a plain proof that over the summits of those mountains the ocean must once have rolled, and this not for a brief space only, but through long periods of time. And not on the mountain-summit only are these fossils known to abound, but sometimes in layer below layer of the mountain, from top to bottom, through thousands of feet of rock.

This may well seem puzzling at first sight. Fossils of sea-creatures on a mountain-top are startling enough; yet hardly so startling as the thought of fossils *inside* that mountain. How could they have found their way thither?

The difficulty soon vanishes, if once we clearly understand that all these thousands of feet of rock were built up slowly, layer after layer, when that portion of the land lay deep under the sea. Thus *each separate layer* of mud or sand or other material became in its turn the *top layer*, and was for the time the floor of the ocean, until further droppings of material out of the waters made a fresh layer, covering up the one below.

While each layer was thus in succession the top layer of the building, and at the same time the floor of the ocean, animals lived and died in the ocean, and their remains sank to the bottom, resting upon the sediment floor. Thousands of such dead remains disappeared, crumbling into fine dust and mingling with the waters, but here and there one was caught captive by the half-liquid mud, and was quickly covered and preserved from decay. And still the building went on, and still layer after layer was placed, till many fossils lay deep down beneath the later-formed layers; and when at length, by slow or quick upheaval of the ground, this sea-bottom became a mountain, the little fossils were buried within the body of that mountain. So wondrously the matter appears to have come about.

Another difficulty with respect to the stratified rocks has to be thought of. All these layers or deposits of gravel, sand, or earth, on the floor of the ocean, would naturally be horizontal—that is, would lie flat, one upon another. In places the ocean-floor might slant, or a crevice or valley or ridge might break the smoothness of the deposit. But though the layers might partake of the slant, though the valley might have to be filled, though the ridge might have to be surmounted, still the general tendency of the waves would be to level the dropping deposits into flat layers.

Then how is it that when we examine the strata of rocks in our neighbourhood, wherever that neighbourhood may be, we do not find them so arranged? Here, it is true, the lines for a space are nearly horizontal, but there, a little way farther on, they are perpendicular; here they are bent, and there curved; here they are slanting, and there crushed and broken.

This only bears out what has been already said about the Book of Geology. It *has* been bent and disturbed, crushed and broken.

Great powers have been at work in this crust of our earth. Continents have been raised, mountains have been upheaved, vast masses of rock have been scattered into fragments. Here or there we may find the layers arranged as they were first laid down; but far more often we discover signs of later disturbance, either slow or sudden, varying from a mere quiet tilting to a violent overturn.

So the Book of Geology is a torn and disorganized volume, not easy to read.

Yet, on the other hand, these very changes which have taken place are a help to the geologist.

It may seem at first sight as if he would have an easier task, if the strata were all left lying just as they were first formed, in smooth level layers, one above another. But if it were so, we could know very little about the lower layers.

We might indeed feel sure, as we do now, that the lowest layers were the oldest and the top layers the newest, and that any fossils found in the lower layers must belong to an age farther back than any fossils found in the upper layers.

So much would be clear. And we might dig also and burrow a little way down, through a few different kinds of rock, where they were not too thick. But that would be all. There our powers would cease.

Now how different. Through the heavings and tiltings of the earth's crust, the lower layers are often pushed quite up to the surface, so that we are able to examine them and their fossils without the least difficulty, and very often without digging underground at all.

You must not suppose that the real order of the rocks is changed by these movements, for generally speaking it is not. The lower kinds are rarely if ever found placed *over* the upper kinds; only the ends of them are seen peeping out above ground.

It is as if you had a pile of copy-books lying flat one upon another, and were to put your finger under the lowest and push it up. All those above would be pushed up also, and perhaps they would slip a little way down, so that you would have a row of *edges* showing side by side, at very much the same height. The arrangement of the copy-books would not be changed, for the lowest would still be the lowest in actual position; but a general tilting or upheaval would have taken place.

Just such a tilting or upheaval has taken place again and again with the rocks forming our earth-crust. The edges of the lower ones often show side by side with those of higher layers.

But geologists know them apart. They are able to tell confidently whether such and such a rock, peeping out at the earth's surface, belongs really to a lower or a higher kind. For there is a certain sort of order followed in the arrangement of the layers all over the earth, and it is well known that some rocks are never found below some other rocks, that certain particular kinds are never placed above certain other kinds. Thus it follows that the fossils found in one description of rock, must be the fossils of animals which lived and died before the animals whose fossil remains are found in another neighbouring rock, just because this last layer was built upon the ocean-floor above and therefore later than the other.

All this is part of the foreign language of geology—part of the piecing and arranging of the torn volume. Many mistakes are made; many blunders are possible: but the mistakes and blunders are being gradually corrected, and certain rules by which to read and understand are becoming more and more clear.

CHAPTER IV.

FIRE-BUILT ROCKS.

'For all those things hath Mine Hand made, and all those things have been, saith the Lord.'—ISA. lxvi. 2.

IT has been already said that Unstratified Rocks are those which have been at some period, whether lately or very long ago, in a liquid state from intense heat, and which have since cooled, either quickly or slowly, crystallizing as they cooled.

Unstratified Rocks may be divided into two distinct classes:

First—Volcanic Rocks, such as lava. These have been cooled at the surface of the earth, or not far below it.

Secondly—Plutonic Rocks, such as granite. These have been cooled deep down in the earth, under heavy pressure.

There is also a class of rocks, called Metamorphic Rocks, including some kinds of marble. These are, strictly speaking, crystalline rocks, and yet they are arranged in something like layers. The word 'metamorphic' simply means 'transformed.' They are believed to have been once stratified rocks, perhaps containing often the remains of animals; but intense heat has later transformed them into crystalline rocks, and the animal remains have almost or quite vanished.

Just as the different kinds of Stratified Rocks are often called Aqueous Rocks, or rocks formed by the action of water—so these different kinds of Unstratified Rocks are often called Igneous Rocks, or rocks formed by the action of fire—the name being taken from the Latin word *ignis, fire*. The Metamorphic Rocks are sometimes described as 'Aqueo-igneous,' since both water and fire helped in the forming of them.

It was at one time believed, as a matter of certainty, that granite and such rocks belonged to a period much farther back than the periods of the stratified rocks. That is to say, it was supposed that fire-action had come first and water-action second; that the fire-made rocks were all formed in very early ages, and that only water-made rocks still continued to be formed. So the name of Primary or First Rocks was given to the granites and other such kinds, and the name of Secondary to all water-built kinds; while those of the third class were called Transition Rocks, because they seemed to be a kind of link or stepping-stone in the change from the First to the Second.

The chief reason for the general belief that fire-built rocks were older than water-built ones was, that the former are as a rule found to lie *lower* than the latter. They form, as it were, the basement of the building, while the top-stories are made of water-built rocks.

Many still believe that there is much truth in the thought. It is most probable, so far as we are able to judge, that the *first-formed* crust all over the earth was of cooled and crystallized material. As these rocks were crumbled and wasted by the ocean, materials would have been supplied for the building-up of layer upon layer.

But this is conjecture. We cannot know with any certainty the course of events so far back in the past. And geologists are now able to state with tolerable confidence that, however old many of the granites may be, yet a large amount of the fire-built rocks are no older than the water-built ones which lie over them.

So by many geologists the names of Primary, Transition, and Secondary Formations are pretty well given up. It has been proposed to give instead to all the crystallized kinds the name of Underlying Rocks.*

But if they really do lie under, how can they possibly be of the same age? One would scarcely venture to suppose, in looking at a building, that the cellars had not been finished before the upper floors.

True. In the first instance doubtless the cellars were first made, then the ground-floor, then the upper stories.

When, however, the house was so built, alterations and improvements might be very widely carried on above and below. While one set of workmen were engaged in remodelling the roof, another set of workmen might be engaged in remodelling the kitchens and first floor, pulling down, propping up, and actually rebuilding parts of the lower walls.

This is precisely what the two great fellow-workmen, Fire and Water, are ever doing in the crust of our earth. And if it be objected that such alterations, too widely undertaken, might result in slips, cracks, and slidings, of ceilings and walls in the upper stories,

* Hypogene Rocks.

I can only say that such catastrophes *have* been the result of underground alterations in that great building, the earth's crust.

For these two leading powers of Nature have been, since the earliest ages of the world's history, perpetually at work, modelling and remodelling, pulling down and building up, each to some extent hindering and yet to some extent helping the other.

By 'powers of Nature' I mean simply powers used by God in nature. For this 'Nature,' of which we hear so much, is but the handiwork of God, the Divine Architect. The powers seen in nature, Fire and Water, Heat and Frost, Gravitation and Electricity, these and a hundred others are His servants, ordained from earliest ages to carry out His will.

We see therefore clearly that, although the first fire-made rocks may very likely date farther back than the first water-made rocks, yet the making of the two kinds has gone on side by side, one below and the other above ground, through all ages up to the present moment.

And just as in the present day water continues its busy work above ground of pulling down and building up, so also fire continues its busy work underground of melting rocks which afterwards cool into new forms, and also of shattering and upheaving parts of the earth-crust.

For there can be no doubt that fiery heat does exist as a mighty power within our earth, though to what extent we are not able to say.

These two fellow-workers in nature have different modes of working. One we can see on all sides, quietly progressing, demolishing land patiently bit by bit, building up land steadily grain by grain. The other, though more commonly hidden from sight, is fierce and tumultuous in character, and shows his power in occasional terrific outbursts.

We in our placid island-home can scarcely realize what the power is of the imprisoned fiery forces underground, though even we are not without some witness of their existence. From time to time even our firm land has been felt to tremble with a thrill from some far-off shock; and even in our country is seen the marvel of scalding water pouring unceasingly from deep underground. Where is the furnace that heats the boiler whence flow those steaming waters in the pleasant town of Bath? What also of the wide hollow, probably an ancient extinct crater, in which the town is built? There have surely been past workings of underground furnace-heat in that neighbourhood, not entirely at an end even now.

But it is when we read about other countries that we better realize the existence of this power.

Think of the tremendous eruptions of Vesuvius, of Etna, of Hecla, of Mauna Loa. Think of whole towns crushed and buried, with their thousands of living inhabitants. Think of rivers of glowing lava streaming up from regions below ground, and pouring along the surface for a distance of forty, fifty, and even sixty miles, as in Iceland and Hawaii. Think of red-hot cinders flung from a volcano-crater to a height of ten thousand feet. Think of lakes of liquid fire in other craters, five hundred to a thousand feet across, huge cauldrons of boiling rock. Think of showers of ashes from the furnace below of yet another, borne so high aloft as to be carried seven hundred miles before they sank to earth again. Think of millions of red-hot stones flung out in one eruption of Vesuvius. Think of a mass of rock, one hundred cubic yards in size, hurled to a distance of eight miles or more out of the crater of Cotopaxi.

Think also of earthquake-shocks felt through twelve hundred miles of country. Think of fierce tremblings and heavings lasting in constant succession through days and weeks of terror. Think of hundreds of miles of land raised several feet in one great upheaval. Think of the earth opening in scores of wide-lipped cracks, to swallow men and beasts. Think of hot mud, boiling water, scalding steam, liquid rock, bursting from such cracks, or pouring from rents in a mountain-side.

Truly these are signs of a state of things in or below the solid crust on which we live, that may make us doubt the absolute security of 'Mother Earth.'

Different explanations have been put forward to explain this seemingly fiery state of things underground.

Until lately the belief was widely held that our earth was one huge globe of liquid fire, with only a slender cooled crust covering her, a few miles in thickness.

This view was supported by the fact that heat is found to increase as men descend into the earth. Measurements of such heat-increase have been taken, both in mines and in borings for wells. The usual rate is about one degree more of heat, of our common thermometer, for every fifty or sixty feet of descent. If this were steadily continued, water would boil at a depth of eight thousand feet below the surface; iron would melt at a depth of twenty-eight miles; while at a depth of forty or fifty miles no known substance upon earth could remain solid.

The force of this proof is, however, weakened by the fact that the rate at which the heat increases differs very much in different places. Also it is now generally supposed that such a tremendous furnace of heat—a furnace nearly eight thousand miles in diameter—could not fail to break up and melt so slight a covering shell.

Many believe, therefore, not that the whole interior of the earth is liquid with heat, but that enormous fire-seas or lakes of melted rock exist here and there, under or in the earth-crust. From these lakes the volcanoes would be fed, and they would be the cause of earthquakes and land-upheavals or land-sinkings. There are strong reasons for supposing that the earth was once a fiery liquid body, and that she has slowly cooled through long ages. Some hold that her centre probably grew solid first from tremendous pressure; that her crust afterwards became gradually cold; and that between the solid crust and the solid inside or 'nucleus,' a sea of melted rock long existed, the remains of which are still to be found in these tremendous fiery reservoirs.

This idea accords well with the fact that large numbers of extinct or dead volcanoes are scattered through many parts of the earth. If the above explanation be the right one, doubtless the fire-seas in the crust extended once upon a time beneath such volcanoes, but have since died out or smouldered low in those parts.

A somewhat curious calculation has been made, to illustrate the different modes of working of these two mighty powers—Fire and Water.

The amount of land swept away each year in mud, and borne to the ocean by the River Ganges, was roughly reckoned, and also the amount of land believed to have been upheaved several feet in the great Chilian earthquake.

It was found that the river, steadily working month by month, would require some four hundred years to carry to the sea the same weight of material, which in one tremendous effort was upheaved by the fiery underground forces.

Yet we must not carry this distinction too far. Fire does not always work suddenly, or water slowly; witness the slow rising and sinking of land in parts of the earth, continuing through centuries; and witness also the effects of great floods and storms.

CHAPTER V

WHAT ROCKS ARE MADE OF.

"God .... which doeth great things and unsearchable; marvellous things without number."—JOB v. 8, 9.

THE crust of the earth is made of rock. But what is rock made of?

Certain leading divisions of rocks have been already considered:

The Water-made Rocks;

The Fire-made Rocks, both Plutonic and Volcanic;

The Water-and-Fire-made Rocks.

The first of these—Water-made Rocks—may be subdivided into three classes. These are,—

    I. FLINT ROCKS;

    II. CLAY ROCKS;

    III. LIME ROCKS.

This is not a book in which it would be wise to go

closely into the mineral nature of rocks. Two or three leading thoughts may, however, be given.

Does it not seem strange that the hard and solid rocks should be in great measure formed of the same substances which form the thin invisible air floating around us?

Yet so it is. There is a certain gas called Oxygen Gas. Without that gas you could not live many minutes. Banish it from the room in which you are sitting, and in a few minutes you will die.

This gas makes up nearly one quarter by weight of the atmosphere round the whole earth.

The same gas plays an important part in the oceans; for more than three-quarters of water is oxygen.

It plays also an important part in rocks; for about half the material of the entire earth's crust is OXYGEN.

Another chief material in rocks is SILICON. This makes up one more quarter of the crust, leaving only one quarter to be accounted for. Silicon mixed with oxygen makes silica or quartz. There are few rocks which have not a large amount of quartz in them. Common flint, sandstones, and the sand of our shores, are made of quartz, and therefore belong to the first class of Silicious or Flint Rocks. Granites and lavas are about one-half quartz. The beautiful stones, amethyst, agate, chalcedony, and jasper, are all different kinds of quartz.

Another chief material in rocks is a white metal called ALUMINIUM. United to oxygen it becomes alumina, the chief substance in clay. Rocks of this kind—such as clays, and also the gems, sapphire, ruby, oriental topaz, oriental emerald, and oriental amethyst, all differing only in colour,—are called Argillaceous Rocks, from the Latin word for clay, and belong to the second class. Such rocks keep fossils well.

Another is CALCIUM. United to oxygen and carbonic acid, it makes carbonate of lime, or pure limestone; therefore all limestones belong to the third class of Calcareous or Lime Rocks.

Other important materials might be mentioned, such as MAGNESIUM, POTASSIUM, SODIUM, IRON, CARBON, SULPHUR, HYDROGEN, CHLORINE, and NITROGEN. These, with many more, not so common, make up the remaining quarter of the earth-crust.

Carbon plays as important a part in animal and vegetable life as silicon in rocks. Carbon is most commonly seen in three distinct forms—as charcoal, as black-lead, and as the pure brilliant diamond. Carbon united in a particular proportion to oxygen forms *carbonic acid*, and carbonic acid united in a particular proportion to lime forms limestone.

HYDROGEN united to oxygen forms water. Each of these two gases is invisible alone, but together they combine to form a liquid.

NITROGEN mingled with oxygen and a small quantity of carbonic acid gas forms our atmosphere.

In the fire-built rocks no remains of animals are found, though in water-built rocks they abound. Water-built rocks are sometimes divided into two classes—those which only contain occasional animal remains, and those which are more or less built up of the skeletons of animals.

There are some exceedingly tiny creatures inhabiting the ocean, called Rhizopods. They live in minute shells, the largest of which may be almost the size of a grain of wheat, but by far the greater number are invisible as shells without a microscope, and merely show as fine dust. The Rhizopods are of different shapes, sometimes round, sometimes spiral, sometimes having only one cell, sometimes having several cells. In the latter case a separate animal lives in each cell. The animal is of the very simplest as well as the smallest kind. It has not even a mouth or a stomach, but can take in food at any part of its body.

These rhizopods live in the oceans in enormous numbers. Tens of millions are ever coming into existence, living out their tiny lives, dying, and sinking to the bottom.

There upon the ocean-floor gather their remains, a heaped-up multitude of minute skeletons or shells, layer forming over layer.

It was long suspected that the white chalk cliffs of England were built up in some such manner as this through past ages. And now at length proof has been found, in the shape of mud dredged up from the ocean-bottom—mud entirely composed of countless multitudes of these little shells, called Foraminifers, dropping there by myriads, and becoming slowly joined together in one mass.

Just so, it is believed, were the white chalk cliffs built—gradually prepared on the ocean-floor, and then slowly or suddenly upheaved, so as to become a part of the dry land.

Think what the enormous numbers must have been of tiny living creatures, out of whose shells 'the wide reaches of white chalk have been made. Chalk cliffs and chalk layers extend from Ireland, through England and France, as far as to the Crimea. In the south of Russia they are said to be six hundred feet thick. Yet one cubic inch of chalk is calculated to hold the remains of more than one million rhizopods. How many countless millions upon millions must have gone to the whole structure! How long must the work of building up have lasted!

These little shells do not always drop softly and evenly to the ocean-floor, to become quietly part of a mass of shells. Sometimes, where the ocean is shallow enough for the waves to have power below, or where land currents can reach, they are washed about, and thrown one against another, and ground into fine powder; and the fine powder becomes in time, through different causes, solid rock.

Limestone is made in another way also. In the warm waters of the South Pacific Ocean there are many islands, large and small, which have been formed in a wonderful manner by tiny living workers. The workers are soft jelly-like creatures, called polyps, who labour together in building up great walls and masses of coral.

They never carry on their work above the surface of the water, for in the air they would die. But the waves break the coral, and heap it up above high-water mark, till at length a small low-lying island is formed.

The waves not only heap up broken coral, but they grind the coral into fine powder, and from this powder limestone rock is made, just as it is from the powdered shells of rhizopods. The material used by the polyps in building the coral is carbonate of lime, which they have the power of gathering out of the water, and the fine coral-powder, sinking to the bottom, makes large quantities of hard limestone. Soft chalk is rarely, if ever, found near the coral islands.

Limestones are formed in the same manner from the grinding up of other sea-shells and fossils, various in kind; the powder becoming gradually united into solid rock.

There is yet another way in which limestone is made, quite different from all these. Sometimes streams of water have a large quantity of lime in them; and these as they flow will drop layers of lime which harden into rock. Or a lime-laden spring, making its way through the roof of an underground cavern, will leave all kinds of fantastic arrangements of limestone wherever its waters can trickle and drip. Such a cavern is called a 'stalactite cave.'

So there are different kinds of fossil rock-making. There may be rocks made of other materials, with fossils simply buried in them. There may be rocks made entirely of fossils, which have gathered in masses as they sank to the sea-bottom, and have there become simply and lightly joined together. There may be rocks made of the ground-up powder of fossils, pressed into a solid substance or united by some other substance.

Rocks are also often formed of whole fossils, or stones, or shells, bound into one by some natural soft cement, which has gathered round them and afterwards grown hard, like the cement which holds together the stones in a wall.

The tiny rhizopods,* which have so large a share in chalk and limestone making, are among the smallest and simplest known kinds of animal life.

There are also some very minute forms of vegetable life, which exist in equally vast numbers, called Diatoms. For a long while they were believed to be living animals, like the rhizopods. Scientific men are now, however, pretty well agreed that they really are only vegetables or plants.

The diatoms have each one a tiny shell or shield, not made of lime like the foraminifers, but of flint. Some think that common flint may be formed of these tiny shells.

* Literal meaning of word, *Rootfoot*.

Again, there is a kind of rock called Mountain Meal, which is entirely made up of the remains of diatoms. Examined under the microscope, thousands of minute flint shields of various shapes are seen. This rock, or earth, is very abundant in many places, and is sometimes used as a polishing powder. In Bohemia there is a layer of it no less than fourteen feet thick. Yet so minute are the shells of which it is composed, that one square inch of rock is said to contain about four thousand millions of them. Each one of these millions is a separate distinct fossil.

One more kind of fossil substance must be touched upon before the close of this chapter, a kind very familiar to all of us from earliest childhood—that of Coal.

What should we all do without coal? How necessary it is to comfort, nay, even to life, in cold countries where an abundance of firewood does not exist! Yet how many use it, day after day and year after year, without a thought of the manner in which it has been provided for them!

Coal comes from underground. Mines, low-sunk, and extending far, in some places even beneath the sea, are made for the purpose of bringing up coal, to be used in thousands of homes. But how did it first get there? That is the question.

Coal is a vegetable substance. The wide coal-fields of Britain and other lands are the *fossil* remains of vast forests.

Long ages ago, as it seems, broad and luxuriant forests flourished over the earth. In many parts generation after generation of trees lived and died and decayed, leaving no trace of their existence, beyond a little layer of black mould, soon to be carried away by wind and water. Coal could only be formed where there were bogs and quagmires.

But in bogs and quagmires, and in shallow lakes of low-lying lands, there were great gatherings of slowly-decaying vegetable remains, trees, plants and ferns all mingling together. Then after a while the low lands would sink and the ocean pouring in would cover them with layers of protecting sand or mud; and sometimes the land would rise again, and fresh forests would spring into life, only to be in their turn overwhelmed anew, and covered by fresh sandy or earthy deposits.

These buried forests lay through the ages following, slowly hardening into the black and shining coal, so useful now to man.

The coal is found thus in thin or thick seams,

with other rock-layers between, telling each its history of centuries long past. In one place no less than sixteen such beds of coal are found, one below another, each divided from the next above and the next underneath by beds of clay or sand or shale. The forests could not have grown in the sea, and the earth-layers could not have been formed on land, therefore many land risings and sinkings must have taken place. Each bed probably tells the tale of a succession of forests.

There are in Great Britain about twelve thousand square miles of coal-fields. In France, in Spain, in Belgium, though coal-fields do exist, the amount is much less. In North America there are about two hundred and eight thousand square miles of coal-fields. Baron Richsofen, who has made very extensive explorations in the Chinese Empire, reports that a large proportion of that country—a vast area, amounting to about one million square miles—consists mainly of coal-deposits. So there is little danger of the supply running short.

CHAPTER VI.

ROCK-LAYERS.

'Before the mountains were brought forth, or ever thou hadst formed the earth and the world, even from everlasting to everlasting Thou art GOD.'—PSA. xc. 2.

BEFORE going on to a sketch of the early ages of the Earth's history—ages stretching back long long before the time of Adam—it is needful to think yet for a little longer about the manner in which that history is written, and the way in which it has to be read.

For the record is one difficult to make out, and its style of expression is often dark and mysterious. There is scarcely any other volume in the great Book of Nature, which the student is so likely to misread as this one. It is very needful, therefore, to hold the conclusions of geologists with a light grasp, guarding each with a " perhaps " or a " may-be." Many an imposing edifice has been built, in geology, upon a rickety foundation which has speedily given way.

In all ages of the world's history up to the present day, rock-making has taken place—fire-made rocks being fashioned underground, and water-made rocks being fashioned above ground though under water.

Also in all ages different kinds of rocks have been fashioned side by side—limestone in one part of the world, sandstone in another, chalk in another, clay in another, and so on. There have, it is true, been ages when one kind seems to have been the *chief* kind—an age of limestone, or an age of chalk. But even then there were doubtless more rock-buildings going on, though not to so great an extent. On the other hand, there may have been ages during which no limestone was made, or no chalk, or no clay. As a general rule, however, the various sorts of rock-building have probably gone on together. This was not so well understood by early geologists as it is now.

The difficulty is often great of disentangling the different strata, and saying which was earlier and which later formed.

Still, by close and careful study of the rocks which compose the earth's crust, a certain kind of order is found to exist, more or less followed out in all parts of the world. *When* each layer was formed in England or in America, the geologist cannot possibly say. He can, however, assert, in either place, that a certain mass of rock was formed before a certain other mass in that same place, even though the two may seem to lie side by side ; for he knows that they were so placed only by upheaval, and that once upon a time the one lay beneath the other.

The geologist can go further. He can often declare that a certain mass of rock in America and a certain mass of rock in England, quite different in kind, were probably built up at about the same time. How long ago that time was he would be rash to attempt to say ; but that the two belong to the same age he has good reason for supposing.

We find rocks piled upon rocks in a certain order, so that we may generally be pretty confident that the lower rocks were first made and the upper rocks the atest built. Further than this, we find in all the said ayers of water-built rocks signs of past life.

As already stated, much of this life was ocean-life, though not all.

Below the sea, as the rock-layers were being formed, bit by bit, of earth dropping from the ocean to the ocean's floor, sea-creatures lived out their lives and died by thousands, to sink to that same floor. Millions passed away, dissolving and leaving no trace behind; but thousands were preserved—shells often, animals sometimes.

Nor was this all. For now and again some part of the sea-bottom was upheaved, slowly or quickly, till it became dry land. On this dry land animals lived again, and thousands of them, too, died, and their bones crumbled into dust. But here and there one was caught in bog or mud, and his remains were preserved till, through lapse of ages, they turned to stone.

Yet again that land would sink, and over it fresh layers were formed by the ocean-waters, with fresh remains of sea-animals buried in with the layers of sand or lime; and once more the sea-bottom would rise, perhaps then to continue as dry land, until the day when man should discover and handle these hidden remains.

Now note a remarkable fact as to these fossils, scattered far and wide through the layers of stratified rock.

In the uppermost and latest built rocks the animals found are the same, in great measure, as those which now exist upon the earth.

Leaving the uppermost rocks, and examining those which lie a little way below, we find a difference. Some are still the same, and others, if not quite the same, are very much like what we have now; but here and there a creature of a different form appears.

Go deeper still, and the kinds of animals change further. Fewer and fewer resemble those which now range the earth; more and more belong to other species.

Descend through layer after layer till we come to rocks built in earliest ages, and not one fossil shall we find precisely the same as one animal living now.

So not only are the rocks built in successive order, stratum after stratum belonging to age after age in the past, but fossil-remains also are found in successive order, kind after kind belonging to past age after age.

Although in the first instance the succession of fossils was understood by means of the succession of rock-layers, yet in the second place the arrangement of rock-layers is made more clear by the means of these very fossils.

A geologist, looking at the rocks in America, can say which there were first-formed, which second-formed, which third-formed. Also, looking at the rocks in England, he can say which there were first-formed, second-formed, third-formed. He would, however, find it very difficult, if not impossible, to say which among any of the American rocks was formed at about the same time as any particular one among the English rocks, were it not for the help afforded him by these fossils.

Just as the regular succession of rock-strata has been gradually learnt, so the regular succession of different fossils is becoming more and more understood. It is now known that some kinds of fossils are always found in the oldest rocks, and in them only; that some kinds are always found in the newest rocks, and in them only; that some fossils are rarely or never found lower than certain layers, that some fossils are rarely or never found higher than certain other layers.

So this fossil arrangement is growing into quite a history of the past. And a geologist, looking at certain rocks, pushed up from underground, in England and in America, can say: 'These are very different kinds of rocks, it is true, and it would be impossible to say how long the building up of the one might have taken place before or after the other. But I see that in both these rocks there are exactly the same kinds of fossil-remains, differing from those in the rocks above and below. I conclude therefore that the two rocks belong to about the same great age in the world's past history, when the same animals were living upon the earth.'

Observing and reasoning thus, geologists have drawn up a general plan or order of strata; and the whole of the vast masses of water-built rocks throughout the world have been arranged in a regular succession of classes, rising step by step from earliest ages up to the present time.

First there are three grand divisions:

I. PALEOZOIC, OR ANCIENT-LIFE ROCKS;
II. MESOZOIC, OR MIDDLE-LIFE ROCKS;
III. CAINOZOIC, OR NEW-LIFE ROCKS.

These, with their chief sub-divisions, are given in a short list at the close of the chapter, beginning with the earliest.

Below the Ancient Rocks lie what some suppose to be the First-formed Rocks, beyond which we know nothing.

In a general way this classification will serve wherever geology is studied. Some geologists arrange differently, but the variations are slight. In certain countries some of the classes or sub-divisions may be entirely lacking, yet the order kept will be the same. The Devonian may be absent, but if present, you will never find it under the Silurian, or above the Carboniferous. The Cretaceous may be missing, but if there, it will not lie beneath the Triassic or over the Miocene. So also with others.

The following list of names should be from time to time carefully referred to, in the course of reading Part II. of this book.

TABLE OF STRATA.

AZOIC, OR NO-LIFE ROCKS.

PRIMARY, OR PALEOZOIC, OR ANCIENT-LIFE ROCKS.

    1. Laurentian.
    2. Cambrian⎫
    3. Silurian   ⎬ *Age of Lower Animals.*
    4. Devonian.—*Age of Fishes.*
    5. Carboniferous.—*Age of Coal.*
    6. Permian.

SECONDARY, OR MESOZOIC, OR MIDDLE-LIFE ROCKS.

    1. Triassic⎫
    2. Jurassic⎬ *Age of Reptiles.*
    3. Cretaceous.—*Age of Chalk.*

TERTIARY, OR CAINOZOIC, OR NEW-LIFE ROCKS.

    1. Eocene *
    2. Miocene †
    3. Pliocene ‡ *Age of Mammals.*
    4. Pleistocene §

Post-Tertiary, or After Tertiary (*Including Age of Man*).

* Literal meaning—" Dawn-New."   † " Less-New."
‡ " More-New."   § " Most-New."

CHAPTER VII.

ROCK-BENDINGS.

"Where wast thou when I laid the foundations of the earth? declare, if thou hast understanding. Who hath laid the measures thereof, if thou knowest? . . . Whereupon are the foundations thereof made to sink? or who laid the corner-stone thereof?"—JOB xxxviii. 4—6. (marg.)

IT is said that the whole of the Stratified Rocks forming the earth's crust, amount altogether to about twenty miles in thickness.

This only means that if you could take each stratum at its thickest part, and arrange it in its place with all the rest above and below, not one of the whole series being left out, the depth of the entire mass would probably be about twenty miles. In the same manner the stratified rocks of Great Britain are said to be over nine miles thick, though in no one spot does their depth approach this amount.

For nowhere in Great Britain, or in all the world, are all the different rocks found together at their greatest thickness. Where one kind is thick, another beneath it is thin, and other rocks are altogether wanting. Sometimes several of the uppermost have either been never formed there, or they have been washed away since formed.

Under New York these rocks are said to be about two miles and a half thick, and in Pennsylvania at least seven miles. On the Continent there are, it is believed, not much less than five miles of stratified rocks, without counting the lower primary layers. So the depth varies greatly in different parts.

But if we realise that these vast masses have been all built up, as it seems, under the ocean, grain by grain, or bit by bit, or shell by shell, of sand or earth or animal remains, we shall find the amount to be wonderful.

These rock-strata give us a 'chain of records' of the past, but the chain is broken in many places, and many links are wanting. At the best, the history is a fragmentary one, with great gaps and sudden jumps, and chasms which the geologist has no power to bridge over except by guesses.

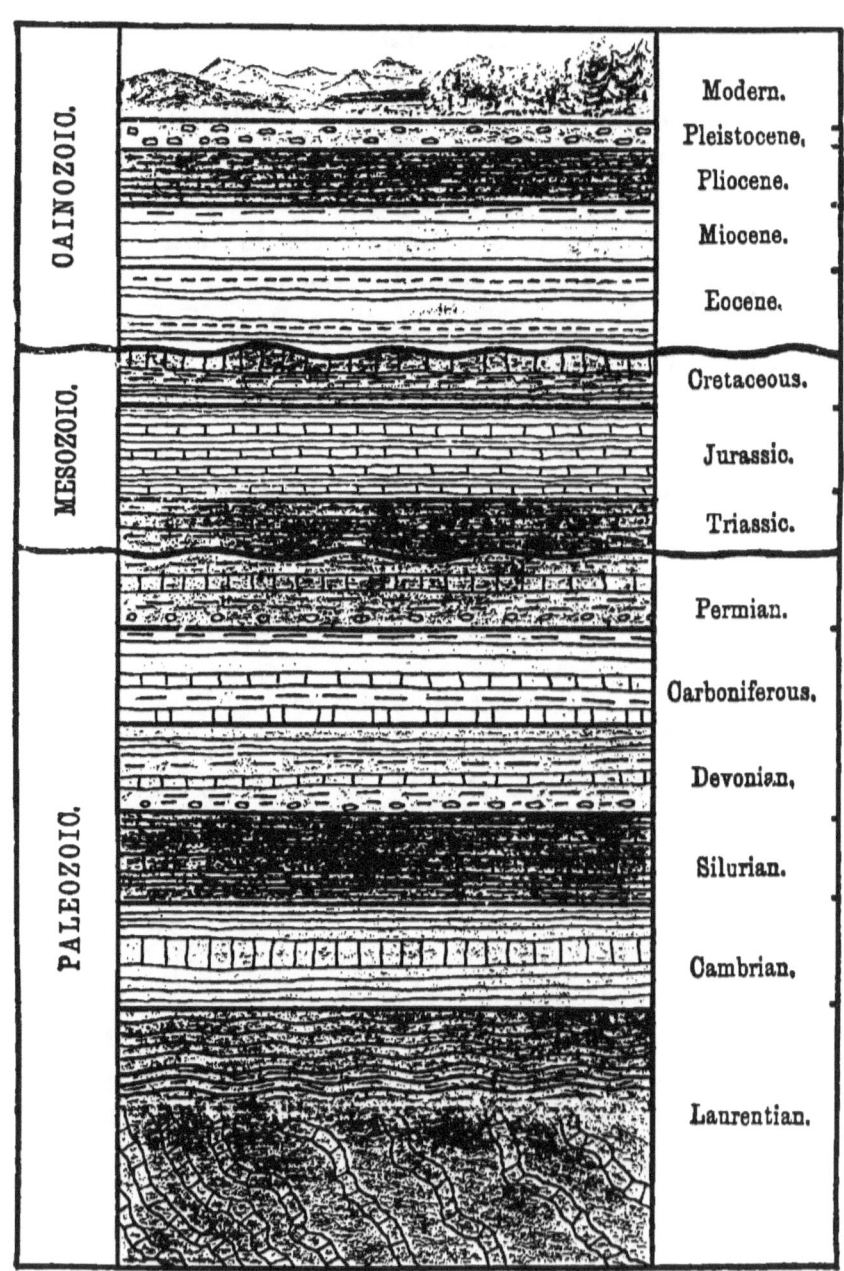

CHART OF STRATA.

One great difficulty in reading geology arises from the fact that it is a volume which is being written chiefly out of our reach. Dry land remains from century to century unchanged, except through the yearly loss of its material, stolen by frost and rain, rivers and sea—and how great this loss is few know. But dry land receives no additions as a rule. All the building up of new land goes on under the ocean.

It would be an immense advantage to the student if he could occasionally take a walk under the sea, so as to be able to note, not only the carting away of material from dry land, but the building up of that material into new land beneath water. For it should be clearly understood and remembered that every yard of fresh mud or rock formed by the sea is made entirely out of material stolen from the land. This double work of taking away from the land and laying down upon the ocean floor goes on incessantly; but only one half of it can we see.

And indeed, even if we could so walk under the sea, like Southey's Kahama, it would avail us little, unless we could watch through long periods of time, with unfailing patience, to see the slow and gradual growth of the building. And were this too possible, we should still stand in need of some power to enable us to look into the far dim past, and to see the precise manner in which this work was then carried on.

For although the manner of rock-making was then probably much the same as it is now, we have no power to speak positively in the matter. It *may* have taken place exactly as it now takes place, or various causes *may* have combined to make both the pulling down and the building up very much more rapid than they are at present. It is easy to conjecture one way or another, but we cannot know which really was the case.

The words 'stratum' and 'layer' have been used often in the last few chapters. Although their meaning is much the same, yet each has its own distinct sense.

A 'stratum' may consist of a great many 'layers.' In a stratum of limestone there are generally several thin layers of limestone one over another, making up the whole mass or stratum. So also a stratum of coal consists of few or many layers of coal. The 'layer' may be either thick or thin, and it may be either loose or joined to the next layer. The stratum' means the whole mass of one kind of rock, lying between two beds or strata of other kinds of rocks—*strata* being the plural of *stratum*.

Suppose you had a great pile of paper, first a number of blue sheets, over them a number of white sheets, then a few pink sheets, over them a number of white sheets again, then a few green sheets. Each sheet would picture a 'layer,' but each supply of one colour lying close together would picture a 'stratum.'

A 'formation' means the whole set of strata believed to belong to one particular age or part of an age in the past. Rocks of different kinds, in different places, if containing the same kinds of vegetable and animal fossils, are said to belong to the same 'formation.'

Rocks are not commonly found, as they were first built up, in flat layers. Sometimes they are bent, curved, slanting; sometimes pushed into an upright position; sometimes for a short space even turned quite upside down, so that a top stratum is seen for a short distance actually underneath a lower and earlier-formed stratum, with plain marks of the catastrophe.

Sometimes it is found that the lower rocks are crumpled, and bent, and distorted, while over these lie some perfectly smooth flat layers. This shows that the disturbance of the lower rocks must have taken place before the upper layers were dropped by the sea. After some great sudden upheaving, or some slow crushing together of the rocks, a calm time followed under the ocean, and sand or earth layers were quietly formed, afterwards not to be bent or broken by the gentle uplifting of that sea-bottom out of the sea.

It seems strange to speak of rocks being bent. But under tremendous pressure even the hardest rocks will yield, curving and folding like soft clay. Also many rocks are known not to be so hard as we commonly see them, when buried underground and sheltered from the air.

Anyone who has been through some of our great iron-working manufactories, and has seen how cold iron can be cut and pierced, twisted and bent, by the deliberate exercise of great force, will wonder less at the effects of pressure on rock. Many of these disturbances past were doubtless calm and gradual, though others must have been sudden and startling— even as now, gentle and slow forces work side by side with fierce and terrific ones. More of this later.

On one of the arms of the Lake of Lucerne, as the tourist goes by steamer towards the little village of Fluellen, he may note a remarkable instance of this rock-bending, in the cliffs to the left. It is in one part as if a gigantic hand had grasped the smoothly lying rock-layers, and had deliberately crushed them into a variety of fantastic crumplings, as a man might crush a piece of paper in his fingers.

Rock-foldings vary in size from tiny bends an inch in depth, to great wrinklings or creasings of the earth's crust, extending through miles of country. Many a range of mountains is an example of such huge wave-like bendings, each crease or wrinkle, divided by a hollow from the next, being thousands of feet in depth.

If no after-change took place, it would be possible to trace in unbroken continuance the lines of the bent layers, running up each mountain and down each hollow. Take half-a-dozen pieces of thick cloth, and fold them backwards and forwards into several massive plaits, to make the matter more clear. Each piece of cloth rising and falling through the bends will picture a layer rising and falling through the mountain-folds.

But in reality the upper layers cannot be so traced, for mountains do not remain unchanged. Frost and rain, wind and torrent, glacier and avalanche, are ever pursuing their work of destruction. The mountains are lower now than they were a hundred years ago, for they are incessantly losing a part of their material.

Loose soil is washed away, stones are carried off, crags and precipices crumble and break, and gradually whole masses are swept from the exposed summits, leaving bare those underlying rocks which still are buried low on the more sheltered mountainsides.

Some chapters back I spoke of the slips and slidings likely to take place in the upper stories of a house, in consequence of lower-story alterations.

Now these slips and slidings have actually taken place in the earth-crust. As the mountains were upheaved, and as low-lying plains were lifted higher— whether by slow or sudden action we cannot always know—there were often slips or displacements in the various rocks disturbed by such movements. These slips are called *faults*, both by geologists and by miners. A 'fault' is found when a mass of rock, containing sometimes several different strata, has slipped down to a lower level than the rocks adjoining, so that each layer is separated from what was once a continuation of itself in the other mass. These 'faults' may be a matter of a few inches or of hundreds of feet. Such a fault in a coal-mine, if extensive, is a serious matter, for the seam of coal will suddenly cease, and will have to be sought for elsewhere. The rocks adjoining having slipped down, in consequence of deep underground disturbances, the continuation of the coal-seam will lie lower also.

CHAPTER VIII.

ICE-WORK.

'He cutteth out rivers among the rocks.'—JOB xxviii. 10.

IN pieces of rock there are found often curious markings, telling each its little tale of bygone days. Sometimes a bit of sandstone is dented over with small round holes or pocks, one here or there breaking into the rim of its next neighbour. These are traces of rain-drops which fell long long ago. Such markings may be seen upon the sand of the sea-shore, commonly remaining only a short time because washed away by the next tide or shower. But in certain cases the sand has been left undisturbed to retain the traces of the last shower which it received in its soft state, before hardening into rock.

Or again a piece of sandstone may be seen to have softly-rounded furrows, running from one end to the other. Here again are water-marks. Did you never note how the retreating tide leaves often upon the sand tiny ridges and hollows side by side, which the next tide smooths away only that fresh ones may be formed? These ridges are caused by the movements of the waves. After a storm they are of a larger size; and it is said that on the Goodwin Sands they are often two feet or more in height. Hardened sandstone sometimes bears these wave-marks, which were imprinted upon it when soft.

Again, stones are sometimes found, curiously marked with lines, as if they had been polished and scratched, not anyhow and in all directions, but in a regular way. Sometimes one set of lines will run across another set, or there will be deep scorings in the midst of delicate even lines, but as a general rule these scratchings all lie straight along the greatest length of the stone. Such stones exist in great numbers, and they were long a serious puzzle to geologists. Even now there is not absolute certainty as to the real cause, though the last and generally believed explanation seems most likely to be the true one.

If you were taking a walk in Scotland, you might be struck with the appearance of certain oft-recurring loose gravel and sand patches, or masses of rugged stones and clay, or heaps of loose rocky *débris*. They do not lie on the higher mountains, but abound in plains and valleys. If you could examine a very little way underground, just below the surface layers of earth or gravel, you would find this unstratified deposit of clay and rough rocky fragments reaching through miles and miles of country; not only in Scotland, but also in many parts of Europe and of North America. It is never, however, found further south than 40 and 50 degrees North Latitude. The same rough unstratified deposit is found in countries of the southern hemisphere, but there in like manner it is not seen further north than 40 and 50 degrees South Latitude. So whatever is the cause of this singular appearance, it plainly has nothing to do with tropical heat.

The deposit is made up chiefly of sand, clay, and stones, sometimes a very stiff kind of clay, with large stones and boulders scattered all through it, and this in Scotland is called *till*. Sometimes there is a layer of such boulder-filled clay, and then a layer of sand, and then more clay, not so coarse and stiff in character. But the clay is never really stratified, and the stones and boulders are always scattered pell-mell through it. Another name by which it is known is *boulder-clay*, and yet another is *drift*. In places it is as much as fifty or a hundred feet in thickness.

The stones and boulders are found as a rule to have been broken off from the rocks and mountains in the neighbourhood. This is easily seen, by comparing the rock of which they are made with the rocks of which the mountains are made.

It is stones found in this clay, and in the heaps of loose *débris*, which are so curiously polished and scratched. They are of every size from tiny pebbles up to huge blocks, and the markings are of every kind from the finest lines to deep-ploughed furrows.

Not only are loose stones thus marked, but large spaces of rock in the same neighbourhoods, bare of earth and sand, will be found to be in a like manner polished and scratched and grooved.

The course which seems to be followed by these stones and rock-markings is somewhat singular. It will run often in the line of larger valleys; but small valleys seem to have been ignored or rather taken at right angles, and hills of moderate size have proved no obstacle.

In addition to the stones and rocks scattered through the 'till,' there are very remarkable blocks of stone often seen, not only on low plains but perched upon high mountains, far away from any other rocks of a like nature. They are called 'erratics' from the erratic way in which they seem to have journeyed across wide tracts of country and crossed deep valleys before reaching their present resting-places. Erratics exist by thousands in the same countries where the drift is found, but they are never seen in the tropics. They too are often polished and scored on one side, like the smaller stones.

On the Jura Mountains of Switzerland the heaps of stones, polished rock-surfaces, and scratched pebbles, are seen in abundance. Also, great rock-masses are found lying there, perfectly different in material from that of the Jura Mountains, but agreeing with that of the Alps some fifty miles distant.

Now, what power could have borne those blocks, some of them as large as a cottage, not only across fifty miles of low grounds, but up the steep sides of the Jura? What power also could have carried these multitudes of rocks and stones, polishing them, scratching them, grooving them, mingling them with clay and earth, and scattering the whole mixture in lavish profusion up-hill and down-hill through hundreds of miles of country?

One of the early names by which the boulder-clay was known was Diluvium or Diluvial Soil. The

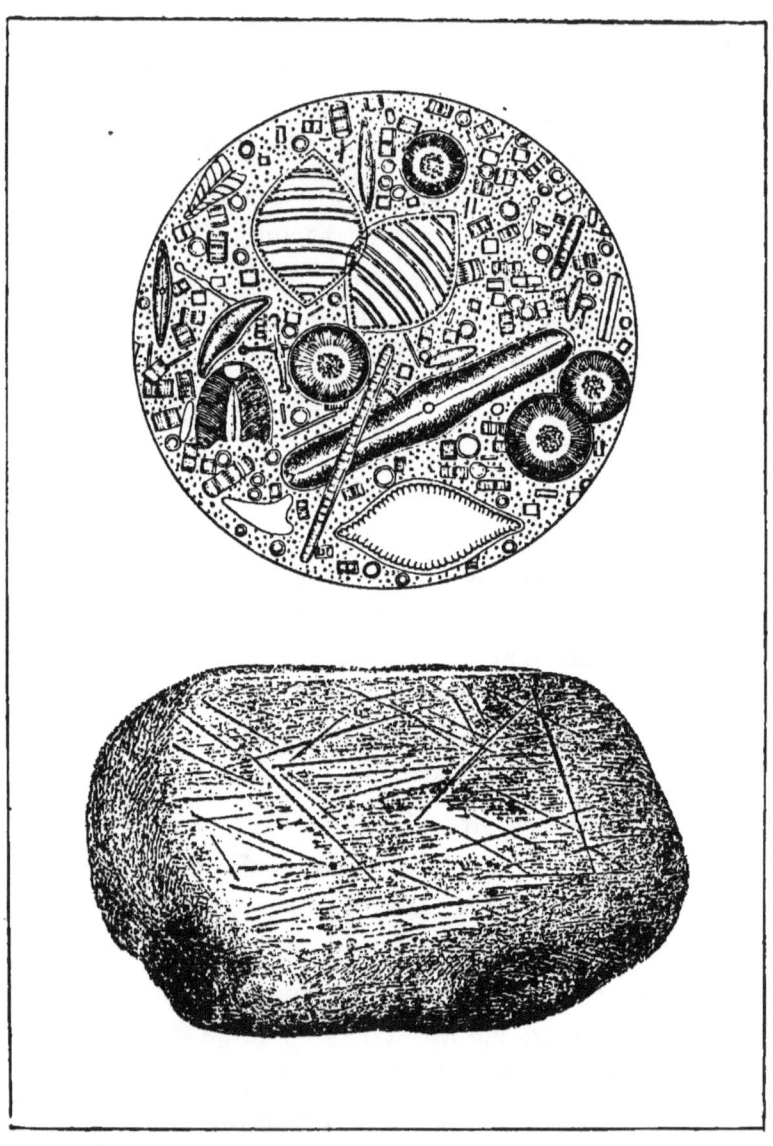

MAGNIFIED SPECK OF EARTH.   SCRATCHED STONE.

Ice-Work. 75

name sprang from the idea that a great flood might have caused the deposit. Some thought that a flood of mud had burst over the continents, bearing rocks and stones before it in confusion. Some spoke of an ocean-flood, and of mighty waves sweeping mud, pebbles, and rock-boulders down valleys and up mountains with resistless force. That water has considerable carrying power is shown by mountain-torrents.

The thought of the flood in the days of Noah rises naturally to mind in connection with this subject. In that vast rush of water over the land, when 'all the fountains of the great deep were broken up,' what might not have been accomplished?

It is well to bear the thought in memory through the study of geology. Great changes probably did then take place; and much of the broken and confused condition of the upper rocks and soils, which adds to the perplexities of the geologist, may be owing to that comparatively recent event.

Still, no mere flood of water, however great, could account for the polishing and marking of thousands of stones and rocks in so regular and systematic a manner. Water will wash stones together, wearing off corners, rounding and rubbing them down, but it will not mark angular stones from end to end with sets of neat lines and grooves. Also, no ocean billows, however mighty, could carry granite boulders across miles of country and lift them up to heights of hundreds and thousands of feet. Both water and mud are utterly incapable of such a task.

In the higher mountains of earth there is a perpetual collection of snow—a vast amount, never melting more than slightly at the surface, and constantly receiving fresh additions.

Now, if the snow had no means of parting with its extra quantities—if it were always taking to itself new supplies and never giving any supplies away, it would increase to an enormous extent. The snow-line, or boundary of perpetual snow, instead of remaining at about the same height year after year, would gradually creep lower and lower down the mountain-sides.

But there are modes of relief for the over-laden mountain. The surface of the snow slowly evaporates or dries away in the sunshine. Every year the summer-heat thaws great quantities of snow in the lower mountain regions, filling rills and torrents to overflowing. Also great masses of loose snow break away from time to time and rush down the mountain, thus lessening the weight above.

Nor are evaporation and thawing and avalanches the only outlets provided.

The great load of snow on the mountain-summit squeezes the lower layers of that snow into solid ice. For ice is of the same nature as snow, only with the particles packed more closely together. The ice thus formed is then pressed out by the same weight from beneath the mass above, and creeps quietly down through the nearest valley. This long tongue of ice, coming from under the snow, and never stopping till it reaches so low down the mountain-side that its further progress is checked by constant melting, is called a Glacier. It is, in fact, a frozen stream—a river of ice. Every lofty mountain which keeps a large amount of snow on its summit all the year round, has its glaciers.

Some glaciers, especially those in the icy regions of the far north, are very large, many miles in length, and miles in width. There are also large glaciers in the higher mountains of Switzerland. A glacier is like a river in many respects, but the quickest motions of a glacier are very slow, compared with the most sluggish of rivers. You might stand for hours by one and never see it move; for the ice creeps along its rocky bed at the rate of only a few inches a day—sometimes as little as one inch, never more than fifty inches. One mile in fifteen years is a very good pace for glacier ice.

Yet the real wonder is, not that these great masses of solid ice should move slowly, but that they should move at all.

As the ice-river crawls onward, its surface is perpetually cracking, and the cracks or chasms or crevasses vary in size from a few inches to hundreds of feet in width and depth. The cracks are commonly across the bed of the stream, not up and down it. They generally come when a sharper slope in the ground puts a strain upon the ice; and later, when the bed becomes more level, the cracks close up and heal or freeze together again.

There are commonly down both sides of a glacier long trails of stones and rock-fragments, which are called *moraines*. Sometimes one or two or more moraines are seen down the middle of the glacier also. Those at the side are fed by fragments falling from the cliffs. Those in the middle are supplied by some other glacier which has joined the first.

Such rock-fragments, if falling into a river, would be ground up into river-detritus—rounded pebbles, sand and mud—and carried out to sea. Falling on the solid ice of the glacier, they are borne slowly down the mountain-side, till the ice reaches the spot

GLACIER.

where it melts, and there they are dropped in a heap called a *terminal moraine*.

Many of the rock-fragments fall into the crevasses, and find their way to the bottom. There, with a vast number of other fragments, torn by the ice from the rocky bed, they are dragged slowly along; and as they move, they scratch and polish the rocks below, and are in their turn scratched and polished by those same rocks. A great many are ground into powder, thus thickening the turbid stream which commonly flows beneath a glacier, during at least part of its course, and rushes out at the terminal moraine.

Numbers of these scratched and scored stones may be found in a terminal moraine; and if a glacier could be removed from its bed, the rocks beneath would be found throughout its course to be marked in a like manner.

It is believed that in olden days, there were glaciers far lower down the mountains, and farther south and wider spread than now—huge glaciers in England, in Scotland, in Switzerland, in France, in other parts of North Europe, and in North America, such as now are only found in ice-bound lands like Greenland.

For the marks on rocks and stones, together with other signs in those countries, seem plainly to agree with the marks of glaciers as seen in the present day.

There was, indeed, another explanation offered, which for a while held sway, before the glacier-theory came into life.

It was agreed that no ocean-waves could lift vast blocks of granite to mountain tops, or could mark thousands of stones as we see them marked.

But what about icebergs? What if once the continents lay deep under water, and icebergs from the frozen north, floating southwards on the sea, dropped fragments of rock and stone here and there on the submerged land, afterwards to be upheaved and to become dry land once more? Might not the same icebergs, grounding here and there in shallower water, scratch and groove the rocks and stones below?

That icebergs do in the present day bear away great pieces of rock and drop them upon the ocean floor is a well-known fact. An iceberg is only a huge fragment broken off from the foot of an enormous glacier; and the rocks and stones lying upon the glacier, or embedded in it below, will in like manner lie upon or cling to the ice-masses, which break off and float away from the glacier. As the iceberg reaches a warmer atmosphere and melts away, such rocks and stones must sink to the bottom of the sea.

But though this may account for some particular

blocks, it will not account for all. Neither will it account for the marked stones. An iceberg grounding might make with its jagged bottom deep scorings and furrows here and there in the underlying rocks, but it could not polish and delicately trace with fine lines thousands of small stones.

We must, therefore, return to the glacier theory, as that which at present seems certainly the most likely explanation.

For in Scotland, and other countries, we have the polished rocks, the scratched and furrowed stones, the piles of moraine-like débris. We have also the huge blocks perched at great heights, where no water could have carried them.

But these glaciers of olden days, if indeed they existed, must have been of enormous extent. They must have covered a great part of Northern Europe. In North America they must have reached to a height of three or four thousand feet up the mountain-side, for the placing of the massive blocks there seen. Scotland must have lain buried beneath one vast glacier-system, branching in all directions.

Sometimes in Switzerland certain smoothed and rounded rocks lie near together, called *roches moutonées*, from their likeness to a sheep's back, and known to have been thus shaped by the grinding power of ice. Such rocks are found in America, where no glaciers now can touch them. The rounded summits of the Scotch mountains would seem to point also to the same cause.

But how could the climate of so many countries have thus differed in the past from our climate of modern days?

This we cannot tell, neither is it needful that we should. Many facts in Nature are plain, which yet we are not able to account for.

A good many explanations have, it is true, been suggested, and a good many theories started—some wildly impossible, some perhaps possible but utterly uncertain. Writers on this subject have, as a rule, been very much more successful in disproving others' notions than in proving their own.

For in truth we simply do not and cannot yet know, either how or when there came about this great period of extreme cold—commonly called the Glacial Age—whether very long ago, or whether comparatively near to historical times; whether only once, or whether several times repeated.

CHAPTER IX.

THE TWO BOOKS.

'Unto Thee, O God, do we give thanks, unto Thee do we give thanks; for that Thy Name is near Thy wondrous works declare.'—PSA. lxxv. 1.

THERE may be, in the past chapters, certain suggestive facts startling to a thoughtful mind, which has not met with them before.

What—a whole world history before the time of Adam? What—whole races of animals coming into existence, living, and dying out upon this earth, before ever the foot of man had touched her soil? What— ages upon ages earlier still of change and growth, of land risings and sinkings, of mountains being built, of rocks being formed, of the earth-crust receiving, touch upon touch, from myriad workers, great and small, her shaping and completion?

Yes, even so. We may not shrink from looking

the matter in the face, for if we rightly read the language of the rocks this is truth, and truth cannot work us harm.

Look at the thousands and tens of thousands of feet of solid rock-strata, reaching through one land after another, all built up—so it seems to us—moment by moment, atom by atom, at the bottom of the ocean. Look at the miles of chalk-formation, composed of countless millions of tiny shells, sinking each one to that same ocean-floor as the little occupant ended its tiny life. Look at the multitude of coral islands in southern seas, formed inch by inch through the lives of ever-busy coral polyps, and note the great masses of limestone rock, made by the grinding of that same coral into powder. Look at the acres upon acres of coal strata, changed remnants of uncountable generations of forest-trees and undergrowth! How long did all these take to be built up and formed and changed, to be raised mountain high, or to be sunk deep underground beneath layer after layer of concealing strata?

How long? Men have rashly sought to answer that question, have vainly endeavoured to count the years of those rolling centuries past. But while the thought of all this is oppressive, is even awful, in the interminable vista of back-reaching ages which it opens out to our dim vision, no living man can say *how long* they lasted. For our rate-measures of growth and change, of rock-building, of limestone formation, of land risings and sinkings, are in themselves variable and utterly uncertain. And even if we could measure and fix their rate in the present, this would not serve us for the past. We can see what goes on before our eyes now. We cannot see or allow for the effects of widely-differing circumstances in the uninhabited earth, during those long-vanished centuries.

But the solemn thought of these ages upon ages brings another question. How about the six short Days of Creation? How about the repeated—'God said . . . . and it was so.'

Stay; the Bible tells us nothing about six *short* days. That is a human addition. Man puts in his explanatory, 'day of twenty-four hours.' The Bible simply says 'Day,' and later on tells us, 'One Day is with the Lord as a thousand years, and a thousand years as one day.' Man writes in margins and in text-books for schools, 'Creation of the World, B.C. 4004.' The Bible, with grand simplicity, says, 'IN THE BEGINNING God created the Heaven and the Earth;' but gives us no clue as to when that Beginning was, or how long the cycles of time might be which lay between it and the creation of man.

'God said ... and it was so.' This does not of necessity imply instant completion. Had God willed He *could* have so created—instantaneously. But from all that we can see, this does not appear to have been His mode of action. 'God said'—the Divine Will acted through the Divine Word—'and it was so'—that which He willed came to pass—'and He saw that it was good.'

We must remember always that we have more Books than one given us by the same Divine Writer. The Bible is the Book of God, but it is not the only Book of God. HE has written for us also the great many-volumed Book of Nature, to which belongs the torn and battered Volume of Geology. The two great Books resemble one another in certain particulars — in being more or less fragmentary in style, and in containing also many matters past our power to understand.

Now it is an absolute certainty that these two Books, both written by the same great Author, cannot in reality contradict one another.

In the study of geology, as of other volumes in the Book of Nature, we can take no safe stand but this. Both are from God. Both are through and through simple truth.

Mistakes may indeed be made. Startling theories and

hasty conclusions may be spread abroad. Our explanations of the one Book may very easily conflict with our readings of the other Book, explanation or reading or both being often utterly mistaken.

But our safety is in a willingness to wait, and to reserve judgment, confident throughout in one broad certainty—that all Truth centres in God, and that what He has written in His Book of Revelation and in His Book of Nature cannot but be in perfect harmony, the one with the other.

There may be grandeur and majesty in the thought that when God created, as He sat enthroned, He spoke the audible word—'Let it be'—and, in answer, worlds, light, seas, land, plants, trees, animals, flashed instantly into perfect being.

But surely the thought is yet more grand, more majestic, and withal more tenderly beautiful, that in the calm unfolding of His power He slowly moulded and fashioned this earth of ours—fire and water, sea and river, tiny polyps and rhizopods, one and all carrying out His will.

Surely the thought contains for us more of sweetness, that through those long preceding ages our Father in heaven was gradually preparing this earth to be our home, was shaping the continents, was restraining the oceans, was causing the mountains to be built and upheaved, was laying stores of fuel deep underground to cheer us in wintry days, was preparing the buried glittering gem to delight our eyes, was making ready the thousand materials which bring comfort and occupation into our lives.

For He thought beforehand on our needs, through all those countless ages. There is much in the past, as in the present, that we cannot fathom, and many a mystery will meet us which we have no power to solve. But one underlying fact is clear. The half-read story of olden days, which I have now to tell, as gathered from the broken records of the earth-crust, is in great measure the story of a Father's loving foresight, a Father's careful preparation for His children's needs.

PART II.

A STORY OF OLDEN DAYS.

CHAPTER X.

TWO KINGDOMS.

'All Thy works shall praise Thee, O Lord.'—PSA. cxlv. 10.

HAVING spent some little time in learning the A B C of the Rock-alphabet, we have now to try to read the story of long-past days, as written in the rocks—or rather, to glance at some few pages in the book.

Before doing so, however, certain words are needful as to different kinds of animals and plants.

Animals and plants are commonly said to belong to separate kingdoms. We talk of the Animal Kingdom, and of the Vegetable Kingdom. All kinds of trees, shrubs, plants, mosses, sea-weeds, are thus called Vegetables.

Both animals and vegetables, in their separate kingdoms, are also divided into classes, some ranking higher and some ranking lower.

It will be seen at a glance that a horse or a dog must stand higher in the Animal Kingdom than a crab or an oyster. It will be seen at a glance that an oak or an apple-tree must stand higher in the Vegetable Kingdom than a moss or a sea-weed.

But between these widely-parted kinds there lie an immense number, descending step by step from the top to the bottom. In Nature we do not find classes divided by empty chasms. Each class or kind of animals or vegetables, instead of being sharply cut off from the next by a wall or moat, passes into it by delicate shades, much as the tints of a rainbow melt softly into one another.

Here we find an animal which belongs clearly to one class, and there we find an animal which belongs clearly to the next class. But between the two we come upon a third animal, which may be said to belong almost equally to either class.

It is the same with the two great kingdoms, Animal and Vegetable. We may decide easily enough that a dog is an animal and that a rose-tree is a vegetable. Yet low down in both classes, on that hazy borderland which joins—not separates—the two, are found certain creatures or things which naturalists were long puzzled to give over to either kingdom. Some were taken for vegetables, and were afterwards found to be animals. Others were taken for animals, and were afterwards decided to be vegetables. Thus kingdom shades softly off into kingdom, kind melts into kind, class glides into class. Hard and sharp dividing-lines seem to be human, not Divine, in their nature.

The same plan is to some extent seen in the great Ages of Geology. We have now to travel in fancy through those long past Ages—shall I call them DAYS?—conjuring up picture after picture. At one time we shall find ourselves in an Age of Fishes; at another in an Age of Reptiles; at another in an Age of Quadrupeds. Yet each age is oftentimes not divided from those before and after by marked barriers, any more than one day is so divided from another day.

There are indeed great gaps in the rock-volume, and many of these Ages seem to have come in and gone out with tumults and disturbances. But it is often impossible to say how far the break is owing to some mighty catastrophe, and how far it is merely caused by many rock-layers having been washed away.

As a general rule, each Age, so far as regards its animals and plants, may be said to have come in and gone out gradually. A few fishes, for example, are discovered in one set of rocks; then in the next set they abound beyond all other kinds of living creatures;

again, in the age following the supply of fishes is found to lessen, and some other creature becomes plentiful in their stead. Thus trees, reptiles, animalcules, four-footed beasts, have each their dawn, their day, their evening decline.

It is somewhat as if we were looking across a succession of lofty mountain-heights, not parted by abrupt precipices one from another, but in most cases by gentle slopes.

Both animals and plants differ from rocks and stones in having *life*, and also in having what are called *organs*. An 'organ' is a certain part of a living body, fitted to perform certain actions. The eye is the 'organ' of sight, and the ear is the 'organ' of hearing. Thus, animals and plants are said to be 'organized,' while rocks are not.

The smaller the number of these separate parts in an animal, the lower the rank of that animal. When an animal has many 'organs' it belongs to one of the upper classes in the kingdom.

The Animal-Kingdom contains every kind of animal which lives, or has ever lived, upon the earth. It is divided into two great lesser kingdoms—one containing all animals which have a backbone, and the other containing all animals which have not a backbone. Each of these two lesser kingdoms is divided again into four classes.*

The very lowest of all animals are exceedingly small and simple in make. Often they cannot be seen at all, without the help of a microscope; and they have no separate parts except a mouth and a stomach. Some have not even these, and are mere tiny bags, able to take in food at any part of their little round bodies. A small dent, or hollow appears, and the food sinks in, after which the mouth vanishes. The tiny rhizopods, from whose shells our chalk-cliffs are made, belong to this class.

The creatures in the next class higher are not quite so simple in make. They have different parts branching out from a centre, something like the petals of a flower, or the spokes of a wheel. Star-fishes and jelly-fishes are of the number.

Thirdly come such animals as the oyster and the snail, and the octopus—soft-bodied in make, with heads and eyes, which is a marked advance on the two lower classes.

Next we reach a very large number of animals, remarkable for their *joints*—not joints in bones, for they have no bones, but joints in a hard outside skin.

* Sometimes the four lower classes are spoken of as four sub-kingdoms.

To this class belong crabs and lobsters, shrimps and worms, spiders and all insects.

So much for boneless creatures. Another step lands us in the higher sub-kingdom,—that of animals with backbones. But here again there are divisions. Cold-blooded animals rank low, and warm-blooded animals rank high. Those that breathe in water rank low, and those that breathe in air rank high.

Beginning with the lowest, the kinds of animals in the second sub-kingdom stand thus:

> I. Fishes.
> II. Reptiles.
> III. Birds.
> IV. Mammals.

There is a kind of small half-class between the Fishes and Reptiles to which the frog and a few other animals belong. Part of his life a frog breathes like a fish, and part of his life he breathes like a reptile.

The great distinguishing mark of the Mammals is that they suckle their young. All four-footed beasts, large and small, from the elephant down to the mouse, are *mammals;* so also are the whale and the seal. Man himself stands at the top of this highest class of all, and belongs to it—for Man is an animal.

But although in his bodily make man is an animal,

yet a great gap lies between him and the highest of the brutes ; since Man alone of all the animal-creation was made in the likeness of the ever-living God.

The Vegetable Kingdom has no sub-kingdoms, but it has two great divisions, which are again parted into classes. The lowest of these divisions contains all *non-flowering* plants, such as sea-weeds, mosses, and ferns. The upper division contains all *flowering* plants and trees.

These few particulars, with the tables following, will serve, through the following chapters, as a slight guide to the order in which Life seems to have entered upon the earth.

Animals and Vegetables were arranged thus in classes, ranging from the lowest to the highest, by zoologists and botanists, quite independently of the discoveries made by geologists in the rocks. It is very remarkable that the order of man's arrangement should be found to agree, so far as it does, with the seeming order of God's creation.

ANIMAL KINGDOM.

Sub-Kingdom I. INVERTEBRATES, or Animals without a back-bone, or inside skeleton.

CLASS

I. *Protozoans*, or *First-Animals:* very small and simple in make; such as tiny microscopic creatures, also sponges, etc.

II. *Radiates:* having parts branching out from a centre, with mouths and stomachs, and sometimes arms or rays; such as star-fishes, jelly-fishes, sea-urchins, etc.

III. *Mollusks:* soft-bodied creatures, with mouths and stomachs, and often distinct heads and eyes; such as the oyster, the snail, the octopus, etc.

IV. *Articulates:* animals having jointed bodies, not joints in bones, but joints in an outside hardened skin, with more organs than any lower kinds; such as crabs and lobsters, shrimps, crayfishes and spiders, wasps, and other insects.

Sub-Kingdom II. VERTEBRATES, or Animals with a backbone and inside skeleton.

CLASS

I. *Fishes:* Cold-blooded, and breathe in water by means of gills.

II. *Reptiles:* Cold-blooded, but air-breathing. *True Reptiles* are always air-breathing. *Amphibians* breathe sometimes in air, sometimes in water; therefore they come mid-way between Fishes and True Reptiles.

III. *Birds:* Warm-blooded and air-breathing

IV. *Mammals:* Warm-blooded and air-breathing; also suckle their young. This is the mark of highest rank among animals. All four-footed beasts, also the whale and seal, also Man, belong to this class.

VEGETABLE KINGDOM.

I. CRYPTOGAMS, or NON-FLOWERING PLANTS.

Characteristics: no flower, and no real fruit. The seed is a tiny *spore*, not a true seed.

First Class (lowest).—To this belong sea-weeds, mushrooms, lichens, and other plants—growing in fronds, without stems.

Second Class (higher).—To this belong mosses and liverworts, and other plants—having short thick stems.

Third Class (highest).—To this belong ferns, and lycopods, and other plants—growing upwards, with longer stems.

II. PHENOGAMS, or FLOWERING PLANTS.

Characteristics: having real flowers and true seeds.

First Class (lowest).—To this belong such plants as the pine and the spruce. They increase in size by growing outward in rings or layers under the bark; also they have a simple flower, and an uncovered seed.

Second Class (higher).—To this belong the elm, the oak, the rose, the apple, and the greater number of trees and plants. They grow like those of the First Class, but have a more perfect flower, and a covered seed.

Third Class.—To this belong the palm, the Indian corn, the lily, and all bulbs and grasses. They do not increase their growth by added rings. This class is not usually ranked as higher than the Second.

CHAPTER XI.

EARLIEST AGES.

'When there were no depths . . . . when there were no fountains abounding with water. Before the mountains were settled . . . while as yet He had not made the earth, nor the fields, nor the highest part of the dust of the world.'—PROV. viii. 24-26.'

GEOLOGY can tell us nothing about the first creation of the world. Properly speaking, geology has only to do with the building up of the earth's crust, after the earth was already made. Still, as the geologist gropes his way through the dim light of past ages, it is not surprising that sometimes we find him wandering on, and trying to discover more about the creation of this round globe.

'IN THE BEGINNING GOD MADE THE . . . . EARTH.'

So much is clear. But how was it made? The

statement is brief, offering no particulars. Geology joins hands with Astronomy in the attempt to explain further. We are brought here to conjecture only. We do not *know*.

Astronomers find in the heavens many solid and cold bodies like our earth ; also others in a burning state surrounded by fiery atmospheres like our sun ; also yet others which seem to be mere masses of flaming gas.

What if these are three stages in the preparation of a world : first, fiery gas ; then, the cooling into liquid form, or into a solid centre with a burning atmosphere around; lastly, the change into a solid and cold globe, more or less ready to be inhabited?

In far-back ages God *may* have made first—if indeed that were the first step—a ball of fiery gas, and sent it rolling through the heavens.

As ages passed, the particles of gas, slowly cooling, would draw closer together, and by-and-by the rolling ball would be a liquid globe, most likely surrounded still for a while by a blazing atmosphere of burning gases, containing much of the materials which now form our rocks. The sun and stars have such atmospheres around them still.

As the cooling went on, a shell of hardening

material would form slowly over the liquid ball, and probably at the same time the centre might be growing solid from the weight upon it. Between the solid inside and the solid outer crust would lie a sea of red-hot liquid, gradually lessening. The underground lakes of molten rock, supposed now to feed our volcanoes, may be the remains of such a vast underground sea.

At first doubtless the crust would often upheave and break, till it grew strong enough to chain down the fiery forces below. A good illustration of this might perhaps be found in the crust seen to form over the fiery lakes of liquid lava, in the volcanoes of the Sandwich Islands.

Then in time, as the crust—this outer crust of first-formed Rocks—grew stiff and unyielding, the cooling of the layers within would cause the inside to become too small for its covering, for cooling bodies always lessen in size. The crust would, therefore, rise and sink in vast ridges and hollows, somewhat like the wrinkling skin of a drying orange.

Meanwhile, from the cooling atmosphere overhead, —no longer a collection of flaming gases, but rather a thick mass of steam or vapour—the entire earth would be gradually overspread by a shoreless ocean of boiling waters. Some of the planets seem to have such dense envelopes of vapour around them still.

Slowly this ocean would gather into the deepening crust-hollows; and low-lying lands would be upheaved beyond reach of the waves, perhaps often to sink and rise and sink again through centuries following. As the waters grew cool enough, the first and lowest kinds of plants and animals might have begun to appear, coming into life mysteriously from the Creator's Hand. And the lessening cloud-masses overhead would admit the first gleams of sunlight and moonlight to the dark waters of the great deep. And already the ocean would have begun its ceaseless work, by wearing away the borders of those low-lying lands, to build up beneath its rolling waves the earliest layers of sediment rocks. How rapidly such work might have taken place under such circumstances we cannot now tell.

Thus, it *may* be, the earliest Ages or Days of creation and preparation passed away.

Some hold the above, or something like it, to be a probable explanation of the course of events. Others are inclined rather to believe that from earliest times our earth was a solid and cool body, and that the oldest fire-made rocks are not of greater age than the oldest water-made rocks. They put quite aside the question of how the earth was created, saying that geology has nothing whatever to do with it, and that geologists had better keep to their particular province—the crust of the earth.

We have, indeed, no means of coming to any decided opinion. We can but hear opposite ideas, and wait for fuller knowledge.

The earliest-formed rocks yet known are chiefly, though not entirely, fire-made rocks.

Although not stratified in the strict sense of the term, they are often arranged in layers, one kind lying over the next, such as granite, slate, limestone, and other kinds, taking turns, one above another. It is by no means certain that the particular rocks classed by some writers as first-formed, were really part of the first-formed crust, even supposing that the crust did actually cool as described above. Some of them, indeed, seem to have been once water-built rocks, afterwards changed by fire into their present state.

When rocks of such ancient date are seen at the surface of the ground, it is because they have been forced up, breaking through all overlying layers; or else because they have not been under the ocean since earliest ages, so that no water-built rocks have been formed over them; or else because, if such rocks ever were formed, the said overlying layers have since been washed clean away.

Few remains of plants or animals are found in these very ancient rocks. There are some dim signs of plants in the oldest of all; and, indeed, this is only what one would expect, apart from fossils, and apart from the Bible record. Vegetables being needful as food for animals, it seems a matter of almost certainty that vegetables must have come into existence first. There are dim signs of animals also in certain limestones of those times. But plants and animals seem to have been of the very lowest and simplest forms.

Can you picture to yourself our beautiful earth in those early days? A wide waste of ocean, over which no ship was ever seen to sail; light breaking dimly through the murky cloud-wrapped atmosphere; a few low-lying lands of bare rocks, with perhaps some lichens here and there, or some seaweeds washed up by the waves; but no plants, no trees, no towns or villages, no sound of man or beast or bird; no mountains and no rivers; no life in the great silent reaches of ocean, save tiny animalcules, passing through their quiet existence. How different from the busy teeming earth of these days!

Only then, as earlier, we may be sure that 'the Spirit of God *brooded* over the face of the waters.'

CHAPTER XII.

THE AGE OF LOWER ANIMALS.

'The sea is His, and He made it; and His Hands formed the dry land.'—PSA. xcv. 5.

WE now come to the great early division of past time, the rocks of which are sometimes called Primary or First Rocks, because they are the first about which we can know anything definite. They are also known as 'Ancient Animal' Rocks,* because none of the animals which lived then were quite the same as any which live now.

Looking back to that dim bygone Period, we seem to see three mountains-tops rise high, one after another, as lofty Alpine peaks are uplifted above all lower summits, with valleys lying between. These three heights, all belonging to the great Ancient Animal Period, are:

* 'Paleozoic means literally 'ancient animals,' or 'ancient life.

The Age of Boneless Lower Animals;
The Age of Fishes;
The Age of Forests, or the Age of Coal.

And first we have to think about the Age of Lower Animals, sometimes called the Age of Limestone-making.

The rocks belonging to this Age extend through a great part of the Continents, but they are generally buried deep beneath other rocks, built over them in later times. It is only here and there that they are to be found, reaching up to the surface of the ground.

There are a good many different kinds of rocks, but the most abundant of all is limestone. It will be remembered that limestone is often formed out of the remains of sea-shells, small or large, ground up and pressed together. This work has to take place under water, and yet not at any great depth; for in deep water the waves have no power to grind together the shells lying at the bottom. Even in the heaviest storms, the bottom of the deep sea is always quite calm.

So two facts seem pretty clear; first, that a great part of our present continents must have been at that time under water; and secondly, that where limestone-making took place, which was over a very wide extent of earth, the water could not have been deep.

Another proof that the water was not deep is that remains of corals are found here and there in the limestone. Reef-coral-making can never go on, either above water or in very deep water. The land—our present land, then under water—was probably slowly sinking through many centuries, before it rose to become dry land, while the coral-polyps were at work building up, and the waves also were at work grinding into fine powder much of the coral, which gradually became hard limestone. This is now seen to take place continually in the islands of the South Seas.

So in the Lower-Animal Age, it seems that our present continents lay for the most part under the sea, though at no great depth. Probably they had already taken something of their present shape, and many mountain-tops—not yet rising high—may have shown as little islands above the water.

Dry land seems to have occupied much less space than at present; but of this we cannot be sure. We do not at all know whether the oceans were then as they are now, or whether what was then dry land has since gone down under the deep sea never to rise again.

The Age of Boneless Animals must—so far as we

are able to judge—have lasted long, though how long no man can tell. Neither is it possible to say when the Age began or when it ceased, counting by years and centuries. The sharp division between period and period is sometimes plainly written in the rocks, but no dates are inscribed there.

Taken generally it seems to have been a quiet time—a time of few disturbances. Long ages appear to have passed away, while the building up of the limestone-rocks went on, layer after layer being silently laid beneath the waves, in shallow seas, with little or no river-mud floating in them. Then some change would take place—perhaps a change in the level of land or sea-bottom—and sand and earth would be borne seaward once more, and here or there sandstone or other rocks would be built up for a while, in the place of limestone, by-and-by ceasing and giving place again to the latter.

So through century after century, slowly or quickly, by the action of water and living creatures and of rising and sinking land, the countries which now we inhabit were being prepared under the sea for our use.

Yet the preparation was not always so quiet and still. Between the Lower and Upper Rocks of the Age, and in a lesser degree at other dates also, marks are seen of great and wide-spread disturbance. Some think it was slow disturbance, lasting long, but this is by no means certain. One way or another, the rock-layers, placed so evenly under the waves, were up-heaved and tilted and bent.

The climate of the earth in those days must have differed not a little from the climate of the earth in these days. It does not appear that there were any strong contrasts of cold and heat. From the most northern parts of Europe and America down to the most southern parts of Europe and the United States, the same animals flourished. Coral-polyps carried on their work not only over submerged England, but far to the north, near the Arctic Circle; and coral-polyps can only exist in warm seas.

In the very earliest of these rocks there are to be seen little ripple-marks and mud-cracks and worm-borings, such as we may see now upon the shore, only hardened with the hardening of the soft sand or clay into rock.

When a piece of rock has such marks, you may be sure that they were made upon the beach between high and low water-line. Is it not wonderful to think that, all those long ages ago, little wavelets played upon the sand, and tiny worms crept to and fro, just as we find now? Only *then* there never came the foot of man, or the feet of little children dancing over the sand, to visit the lonely shores.

But the world was not so empty at that time as in the earlier Ages of the last chapter. Life had already begun to abound, though it was as yet only lower life in kind.

With regard to the land-plants, we can say so little as to be almost nothing. Many land-plants *may* have grown on countries or islands which *may* have sunk afterwards down below the sea. But if any such there were, we have no trace of them. With regard to ocean-plants, we only know that there were sea-weeds; and sea-weeds belong to the very lowest order of the flowerless and seedless division of plants.

Limestone keeps animal remains well, but it is rarely known to keep any vegetable remains. This may explain in part the scarcity of stony vegetable remains in these rocks, limestone having been the most abundant.

But if vegetable-fossils are rare, animal-fossils are not. The seas swarmed with living creatures; only there were none of the higher ranks. No beasts, no birds, no reptiles, and, until the very close of the Age, no fishes either, seem to have existed. Back-boned animals were as yet uncreated. The boneless lower animals had the earth to themselves in those days, and their fossils are found in great numbers.

The ocean must indeed have teemed with life. Yet of all the living creatures which ranged the seas, not one was precisely the same as any one living creature of the present time. Some bore a certain degree of resemblance to modern kinds; others were utterly unlike any animal now seen. Nay, the very creatures of the first half of the Age differed from the creatures of the last half, and those again from the creatures of the Age following.

How the change came about we cannot tell. We can only see that there was a break, and that the animals which lived after the break were not the same as the animals which lived before the break. It seems to be the close of one chapter in the record and the beginning of another chapter. But whether this really is the case, or whether the seeming break is only caused by some leaves in the book having been torn out and lost, is a difficult question. Some suppose the one and some suppose the other, and no one really knows.

All we know is, that the older kinds of sea-creatures did die out, and that the newer kinds—very much like the old, yet with marked differences—did come in. The Hand of God was working through those great Creation-Days, but precisely *how* He worked we are not able to say.

The animals living in this Age belonged to all the four lower classes in the Animal Kingdom. No land-animals' remains have been found. They were all sea-creatures. The ocean seems to have been thickly populated; and limestone-making must have advanced grandly in those busy waters, where no larger creatures lived, as now, to prey upon the helpless multitudes.

The fossil remains of one curious three-lobed animal have been found in great abundance. The name 'Trilobite' has been given to it, from the shape; but no man has ever seen the living creature, for nothing like it is in the earth now. Trilobites flourished all through the Age which we are now thinking about, little ones and big ones, sometimes as much as a foot in length, with large well-marked eyes. These eyes, visible in the stony remains to the present day, tell a tale of interest. Even in very early ages there is supposed to have been considerable light, or the trilobite might have found his eyes useless to him. So we may learn something from even so simple a matter.

But the trilobite is one of those 'ancient animals'

which have died completely out of existence. As has been said of the ammonite:

> 'The Almighty's breath spake out in death,
> And the *trilobite* lived no more.'

Towards the end of the Age a few remains of land-plants—flowerless ones—are found for the first time.

Another and still greater event is the sudden appearance of Fishes. Up to this date, all animals are of the lower classes, boneless in make. All at once a change takes place, a marked advance is made, and Fishes burst upon the scene. They come abruptly, without warning,—not gradually introduced, step by step,—not found at the beginning of a new Period, after a mysterious gap in the written record,—but breaking suddenly upon us, full-blown creatures of a new and higher order, before the close of the great Lower-Animal Age.

Until the year 1859 the oldest fish-fossils known were those in a 'bone-bed' of the latest rocks belonging to this Age.

The said 'bone-bed,' strewn thickly with fossil remains of fishes, is in Scotland, reaching through about a hundred miles of country. Eye-witnesses describe it as literally crowded with fish-skeletons—bones and teeth, spines and scales, all mingled together, all transformed into jetty stone.

The appearance of these dead creatures is remarkable in another way. There is every sign about them of a sudden and violent death—bent and contorted figures, outspread fins, extended spines, tails sometimes curved round so as to meet the head. They seem to have passed by hundreds out of existence in sudden agony and fear, yet scales and spines and rays of delicate make are found in multitudes uncrushed and uninjured. What manner of death overtook them we can only conjecture, but a clean sweep appears to have been made of this early fish-generation. No more such remains are found for some distance in the rocks overlying.

Since the date above mentioned, discovery has been made of one earlier specimen in the shape of a fossil fish nearly allied to the modern sturgeon. This is remarkable, since the sturgeon does not rank as one of the lower orders of fishes.

It would be very rash to declare that this or any other particular fish was the first of its kind which ever lived—in other words, that it was the first backboned animal created. More such remains, further back in the record, may no doubt be from time to time found.

But the rocks of this Age have been perhaps more diligently and widely studied than any others, and amid the hundreds of earlier animal-remains brought to light, no earlier fish-fossils have appeared. It is scarcely possible that any later discoveries should do away with the remarkable manner in which the Fish bursts upon us. There can be little doubt that, until near the end of this great early Period, fishes *were not*, and then—suddenly, so far as our knowledge is concerned—then, fishes *were*.

'God said, Let there be . . . . and it was so.'

CHAPTER XIII.

THE AGE OF FISHES.

'Who laid the foundations of the earth, that it should not be removed for ever.'—PSA. civ. 5.

THE rocks lying next above in order, tell us of the great Age of Fishes, following close after the Age of Lower Animals. Another leaf of the book has to be turned, and another chapter has to be entered upon.

A goodly part of the rocks, now to be considered, are known as the Old Red Sandstone, because they are formed of sandstone more or less reddish in hue.

Above all these rock-strata of the Fish Age, lie the great Coal Formations, and they again are surmounted by more red-tinted sandy rocks, to which the name of New Red Sandstone has been given.

A large amount of sandstone belongs to the Fish

Age, and also a large amount of limestone, abounding in remains of corals.

Where sandstone is found, there are no coral-remains; but where limestone is found, there the little coral-animals have evidently worked hard. At the Falls of the Ohio in America, great piles of coral of the Fish Age may be seen, just as they were long ago heaped together by ocean waves. In English limestone also of that period coral-remains exist.

These facts seem to point once more to the probability of a generally milder climate in the earth than we now enjoy. Also it seems that a great part of our present continents must still have been under water, overflowed by shallow seas. While coral-building went on, there was most likely land-sinking; but this may have been only temporary. The general tendency of the continents would rather have been to rise.

It should be kept in mind that, all this while, the land upon which we now live was being slowly built under water. Portions indeed were already heaved up beyond reach of the waves, sometimes to remain thus undisturbed until the present time, sometimes to rise and sink repeatedly, before becoming settled dry land. But as yet the building was far from finished.

The lowest rocks, or cellars of the house, were done.

The first water-built rocks, or the basement-rooms, were also done. The second series of water-built rocks, or the ground-floor rooms, were now in hand. Grain by grain of sand dropping through the restless waters; inch by inch of coral, built by the coral-polyps and ground together by the waves;—thus the work went on.

But note the calm forethought, the Fatherly care, evidenced by all this. Little knew the coral-polyps for whom they were so busily employed—as little as guessed the grains of sand wherefore they were swept to and fro to settle down upon the ocean-bed. Does the brick in the workman's hand know or care why it is placed in the wall? In either case we see the control of mind—we see thought, intention, watchfulness.

How much of land was in the earth at that date we cannot tell. Islands or continents then existing may since have sunk beneath the waves; on the other hand, there may have been only the beginning of our present continents and islands, the higher parts showing above water, the lower parts still overflowed. We only know with certainty that much of our present dry land was not then dry land.

Beside wide reaches of coral-seas, and broad expanses where droppings of sand-grains went quietly on through the centuries, there were also sandy and muddy flats, lying between high and low water line, where waves and sunbeams could play alternately. This is shown by ripple-marks and sun-cracks, hardened in the sandstone of that period.

A marked difference is to be seen between the living creatures of this Age and the Age before.

For the last was distinctly the Age of animals without a back-bone or inside skeleton. This was distinctly an Age of animals with a back-bone and an inside skeleton.

Fishes do, it is true, belong to quite the lowest class in the great upper division of back-boned animals; nevertheless they are of a rank markedly higher than any creature belonging to the lower division of the Animal Kingdom.

In plants, as well as in animals, there was advance beyond the former age. Seaweeds seem no longer to have stood almost alone. Ferns, and club-mosses, and other members of the highest class of Flowerless Plants, as well as conifers, which belong to the lowest class of Flowering Plants, grew in plenty. Each tree and plant was indeed unlike any tree or plant ever seen now; still there was just enough likeness for botanists to be able to name many of them.

The bare rocks had a green clothing at last. No

oaks, no elms, no beeches, no fruit-trees, graced those early forests, yet they were not lacking in beauty. Many kinds of tall and graceful ferns grew there; and huge tree-ferns; and great cone-bearing trees like pines; and yet other ancient trees, with large trunks thirty feet high, and long palm-leaf fronds waving in a crown at the top.

No four-footed creatures ranged through the forests, and not a sign has been found of bird, or even of reptile. Yet the woods, silent but for the murmuring of the wind and the rustling of the fronds, were probably enlivened by the hum and buzz of many an insect— such insects as we do not see in these later days.

For the earliest-known insect-remains have been found in the rocks of the Fish Age, being remains not only of the earliest-known insects, but of the earliest-known land-animals of any kind. It is remarkable that hitherto we have had to speak of sea-life only. One of these insect-fossils is of a huge May-fly, no less than five inches across from tip to tip of its wings. If such May-flies darted to and fro in the air, what manner of spiders and ants and beetles *may* not at the same time have crept over the ground?

But no four-footed beasts were yet created, and no reptiles. Not even the croak of a frog seems to have mingled with the buzz of insects in the woods.

With regard to birds we can say little. It is remarkable how few remains of birds have been found, even in rocks of much later date, belonging to periods when they are believed to have abounded.

We must remember that these fossils of animals have been almost always preserved, in the first instance, under water. Now a bird is of very light make—so light that if he dies and falls into the sea, his body will probably float until it decays or is devoured. Certain very large birds, such as ostriches, would form an exception; but this explains why so very few of the smaller birds are found as fossils. It is impossible for us to say, from the rock-records, at what time birds were probably first created.

When the word 'time' is used thus in Geology, it should be understood as reckoned, not by years and centuries, but by the successive rock-layers. The Geologic Ages cannot be counted in years, for the rock-layers supply us with no dates, and guesses are worthless.

If few living creatures as yet inhabited the land, it was not thus with the sea. The ocean teemed with life, not only as it had done in the past, but even more abundantly, for this was the great Age of Fishes, and of fish-fossils in the rocks there is indeed a rich supply. The number of British Species alone—that is, the number of the different kinds of fishes which swam about in the shallow seas overflowing a great part of England and Scotland—amounts to more than one hundred.

There were a good many sharks, but the greater proportion of fishes were of a kind remarkable for their shining scales. In the present day there are only a few belonging to this family, and they are only found in some rivers of North Africa and North America. In those times there were very large numbers belonging to this family, and they flourished in British seas as well as elsewhere.

It is a question whether many of them were not fresh-water fish; if so, Britain must by that time have risen so far out of the ocean as to have been no longer overflowed by the salt waves; although wide shallow lakes must have been spread over considerable parts of England and Scotland.

In addition to fishes, there were plenty of corals and sponges; also a new kind of large trilobite; also enormous lobsters, from four to six feet in length. These last would be somewhat unpleasant to meet upon the seashore. The great King Crab of China, three feet in length, pales before the gigantic lobster of olden times.

CHAPTER XIV.

THE AGE OF COAL.

'O Lord, how manifold are Thy works! in wisdom hast Thou made them all: the earth is full of Thy riches.'—PSA. civ. 24.

IN the last chapter I spoke of the lower stories of the great Earth-crust building as completed—cellars, kitchens, basement, and ground-floor. But, on second thoughts, this was surely a mistake. The building has not yet advanced so far above the foundations. We have but reached, as it were, the vast cellars of the house, wherein our Heavenly Father, with thoughtful care for future generations of men, has laid up an exhaustless store of coal.

When the great Age or Day of Coal-preparation rolled in upon the Earth, there were as yet no coal-seams hidden in her crust, no wide-reaching coal-fields in Europe or America. Sandstone, and limestone, and other rocks were built one upon another, but of coal-strata none could have been discovered by the most patient mining.

We find ourselves, in the Coal Age, upon a world, no longer of bare rocks and dreary fungus, no longer of few islands and scanty woods. The continents are rising out of the ocean, and also out of the broad inland seas; not indeed in a manner of unbroken advance, but rather as the tide comes in, gaining ground upon the whole, though often seeming to lose it. Just so, with many risings and sinkings, but with a general upward movement upon the whole, low extensive lands have emerged from the waters. And from shore to shore these lands are densely covered with a wild luxuriance of forest-growth. The growth, it is true, differs greatly from forests of modern days, yet it is by no means lacking in grandeur and beauty.

Let us try to picture to ourselves what kind of scenery was to be found upon earth in those days. A painting of one part would be a painting of ten thousand parts, for the same forests ranged north, south, east, and west, in Europe and America, in the tropics and in the Arctic zone. One kind of climate seems, at least in a great measure, to have overspread the earth.

Come with me,* then, in fancy, leaving the present far behind us, back to those early ages, and stand with me upon some low ancient hill, which overlooks the flat and swampy lands of the American continent. Few heights of any magnitude are yet to be seen. The future Rocky Mountains lie still beneath the surface of the sea, and are the scene of slow and busy limestone-making—one of God's workshops, where myriads of workers carry out His will. The Alleghanies are not yet heaved up above the level surface of the ground, for over them are spread the boggy lands and thick forests of future coal-fields. The Mississippi is not yet in existence, or if in existence, is but an unimportant little river; for these low lands can supply no such mighty volume of water as now rolls into the Mexican Gulf.

Below us, as we stand, we can see a broad and sluggish stream creeping seawards, widening ofttimes into shallow lakes. And either side of the stream, on broad marshes and swampy plains, vast forests extend in every direction as far as the horizon, bounded on one side by the distant ocean, clothing each hilly rise, and in rich luxuriance sending islets

---

* 'I must confess my debt, in this little description, to a paragraph of vivid word-painting in 'Analogies of Nature and Grace,' by the Rev. Professor Pritchard, D.D.

of matted trees and shrubs—such as are seen in the present day in tropic countries—floating down the waters.

Strange forests these to our unaccustomed eyes. No oaks or elms, no beeches or birches, no planes or sycamores, no palms or many-coloured wild-flowers are there; but enormous club-mosses, and gigantic horse-tails, and splendid pines, and abundance of ancient trees with wide waving frond-like leaves, and graceful tree-ferns, and countless ferns of lower growth filling up all gaps.

No wild quadrupeds of the higher grades yet breathe in any quarter of the earth, and the silent forests are enlivened only by the stirring of the breeze among the trees and the occasional hum of large insects darting to and fro. But upon the margin of yonder stream a huge four-footed creature, like a gigantic salamander, creeps slowly along, leaving the impress of his broad soft feet in the yielding mud, while a gleam of light plays upon the hard and shiny scales in which his body is clothed, as he slinks under the herbage overhanging the water and disappears.

Only a gleam of light, not sunshine, for little or no sunshine can creep through the misty atmosphere. The earth seems clothed in a garment of clouds, and

Forest of the Carboniferous Age.

the air is positively reeking with damp oppressive warmth, like the air of a hot-house. This explains the luxuriant growth of foliage.

Could we thus stand upon the hill-tops, and keep watch through the long Coal-building ages, we should see generation after generation of forest-trees and undergrowth, living, withering, dying, falling to earth. Slowly a layer of dead and decaying vegetation thus collects, over which the forest flourishes still ; tree for tree, and shrub for shrub, springing up in the place of each one that dies.

Then after a while, through the working of the mighty underground forces, the broad lands sink a little way—perhaps only a few feet—and the rushing ocean-tide pours in, overwhelming the forests, trees and plants and living creatures, in one dire desolation. Nay, not dire, for the ruin is not objectless or needless. It is all a part of the same wonderful foreseeing preparation for the life of man upon earth.

Under the waves lie the overwhelmed forests, prostrate trunks and broken stumps in countless numbers overspreading the gathered vegetable remains of centuries before. Upon these the sea builds a protecting covering of sand or mud, more or less thick, and sea-creatures live there, and fishes swim hungrily to and fro, and the latest generations of trilobites die and find a sepulchre in the unfinished layers, by-and-by to become firm rock, with stony animal remains embedded in it.

After a while—how long a while we have no means of even guessing—the land rises again to its former position. Bare sandy flats once more extend, as in former days, but they do not long remain bare. Lichens and hardier plants soon find a home, and earth gradually collects—possibly brought by river-floods—and the light spores of those ancient forest-trees, easily carried by the wind, take root and grow, and luxuriant forests, like to the last, spring anew into being. Again the wide stream bears upon its waters islets of matted plants and trees; and upon river and lake-bottoms and over the low damp lands rich layers of decaying vegetation again collect, increasing through the centuries. Then once more the land is seen to sink, and the ocean-tide pours in, and another sandy or muddy stratum is built up on the overflowed lands. Thus the second layer of forest-growth is buried like the first, and both lie quietly through the long ages following, hidden from sight, slowly changing in their substance from wood to shining coal.

Thus time after time, here twice or thrice, there far more frequently, the land rose and sank, rose and sank, again and again, that supplies of coal might be laid up in earth's store-houses for our use. Not that a whole continent is believed to have risen or sunk at once; but here at one period, there at another period, the movements probably went on.

How rapidly the growth, death, and decay of trees and plants may have taken place in those early times, under circumstances so different from our own, we have no power to say. Also we are utterly in the dark as to whether the land rose and sank slowly through long periods of time, or whether the movements came about suddenly through mighty heavings from below. Some facts would seem to point to one conclusion, some to the other.

The seams of coal, found between other rock-strata, vary much. Some are scarcely thicker than a piece of paper, while others are as much as thirty or forty feet in thickness. One to ten feet is, however, more usual.

When we consider the amount of forest-trees which would be needed, when decayed, pressed together, and hardened, to form even one foot of coal, the matter is sufficiently bewildering. We can scarcely venture to picture to ourselves the length of time needed, according to our present notions and experience, to form a layer of vegetation thick enough to be transformed into thirty or forty feet of coal. Nor is it needful that we should, since our rates of measurement as to the growth and decay of plants are useless to explain what *may* have been the rate of their growth and decay in long-past ages.

Beneath each layer of coal is commonly found a layer of clay, varying from a few inches to many feet in thickness. This clay is sometimes called 'fire-clay,' because out of it bricks can be made which will stand fire.

For a long while people were puzzled by these clay layers. They are seen to be more or less discoloured, as if liquid from the decaying trees and plants had soaked into them. It is now believed that the clays were once the soils of the ground, upon which the forests grew, one such soil to each forest being shown by one layer of clay under each layer of coal.

Some very curious vegetable remains were found in the clays, and were supposed to be a kind of branching water-plant. The name Stigmaria was given to the plant. But after a while it was discovered that instead of being a separate plant it was really only the branching root of the tree Sigillaria— already spoken of as an ancient tree with waving frond-like leaves. The trunk and the root were found joined together in a way that could not be mistaken.

Many other roots in these clay layers were once taken for water-plants.

So the clay was the forest-soil, and the roots of the forest-trees grew in that soil, while the trees themselves flourished above.

The greater mass of vegetable remains, making up the coal, decayed slowly; but when the final ruin of the forest came, whole trunks were snapped off close to the roots and flung down. These are now found in numbers on the tops of the coal-layers, the barks being flattened and changed to shining black coal.

Sometimes a trunk, instead of falling, kept its upright position under water, and was built up in that manner by the sand layers. The inside of the tree-trunk decayed, and sand gradually filled it, while the bark changed to coal. Such trunks are common in coal-mines, and are much dreaded by the miners, for they often fall suddenly, and kill anyone that happens to be below. The coal-bark cannot support the heavy sandstone inside, when other supports are removed by the working of the mine.

These tree-trunks, as above said, are often upright. This only means that they are upright as regards the coal-layers in which they grew. If the coal-layer is flat, the tree-trunk will still point upwards. But if, as is more usual, the coal-layer has been heaved into a slanting or nearly upright position, then the tree-trunk will slant, and may even lie flat across the roof of the mine, though its position really is *upright from the coal.* The danger to miners of such trees falling from the roof may be readily understood.

How wonderful the tale of olden days told to us by these buried forests, by the sand-filled tree-trunks, by the fossil leaves and stems in coal, by the old clay forest-soils, and by the ancient tree-roots still remaining in those soils!

CHAPTER XV.

MORE ABOUT THE AGE OF COAL.

'Many, O Lord my God, are Thy wonderful works which Thou hast done.'—PSA. xl. 5.

ALTHOUGH the thought of Coal is put strongly forward, in speaking of this particular Age, because it is of chief importance to man; yet the actual amount of coal is small, compared with the amount of other kinds of rock.

For instance, the coal-beds in South Wales are about ten thousand feet thick. But the thickness of the actual coal-seams, if put all together, would be only about one hundred and twenty feet.

In one place near Swansea there are coal-rocks more than three thousand feet thick. Sixteen separate masses of sandstone lie in these rocks, one alone being five hundred feet in thickness. Layers of other rocks also part the sandstone layers. The coal-seams or actual coal-layers, lying between the rocks, are sixteen in number, some only one foot thick, some more,—one as much as nine feet. Yet these slender black bands tell of no less than sixteen distinct forests, each of which in turn sprang into being, flourished through centuries, and was overwhelmed. The tale is amazing; but in other places the number of buried forests is far greater.

In the Coal-seams abundant remains of plants and trees are found, also some remains of land animals. In limestone of the same period there are fossils of sea-shells, corals, and other salt-water animals.

The coal-layers thus tell of forests on land; the limestone tells of overflowing ocean-water; and the sandstone may tell of either salt or fresh water. When coal and sandstone and limestone are found built up alternately, one layer over another, we are reading of strange changes in the past—of a land at one time dry, or at least boggy, at another time under the sea, at another time perhaps under a fresh-water lake.

These rocks of the Coal Period are all stratified. In any piece of coal you may see the thin even layers. They show distinctly, but do not separate easily.

The climate of earth in the Age of Coal is believed to have been warm and moist, with heavy mists and little sunshine. This would favour the rich growth of such plants as then flourished.

The abundance of ferns seems to tell of a damp atmosphere, and the great extent of forests over all parts of the earth, made up of the same kinds of trees, whether in the tropics or in the far north, seems to tell of a very equal climate, little hotter in one place than in another. It has been suggested that the cause of this prevailing mildness may have been the greater internal heat of the earth, in those days; but about this we cannot speak positively, any more than about the cause of the intense cold which seems to have reigned over the earth much later. There is, however, little doubt as to the fact of this general warmth. Corals, which now are only to be found in warm southern seas, then lived in the Arctic Ocean. Vegetation now is quite different in different parts of the world, but then the same trees grew in Greenland and in France.

Although the climate was probably warm and damp, it is not supposed to have been intensely hot. In a tropical climate vegetation decays very rapidly, and this would not agree with coal-making, which requires slow decay and large deposits of vegetable remains. It was probably warm and soft and moist, with no extremes of either heat or cold.

Some believe that a very much larger amount of Carbonic Acid Gas was in the atmosphere then than now, but others doubt this.

Great numbers of vegetable fossils are found in coal—leaf and fern impressions, seeds, branches, barks, and trunks. Most of the plants and trees were flowerless in kind. Possibly the fact that they had not for the most part regular seeds, but tiny light spores, easily borne upon the breeze, may help to explain the vast extent of the forests.

Some hundreds of different kinds of ferns, club-mosses and horse-tails, are known to have grown. They seem to have been chiefly remarkable for their great size, as compared with such plants of the present day. *Now* a forest of non-flowering plants would be scarcely higher than a man's head, but things were different then.

Tree-ferns, which in these days grow only in warm countries, in those days flourished over all the earth. Club-mosses, which now are never seen more than four or five feet high, then grew to a stature of sixty or seventy feet. The horse-tails of the present stand at the most scarcely three feet above ground, and are hollow-stemmed; but their nearest relatives of the Coal Period had woody trunks as much as twenty feet in height.

FERN FOSSILS IN COAL.

The ancient tree 'Sigillaria,' now never seen, was then abundant. Its fossil was at first taken for that of a fern, until it was found to have had long leaves unlike that of any known fern. The trunk often reached a height of sixty feet or more, being bare, except at the top, where the long frond-like leaves grew. This tree belonged to the lower division of plants, but it seems to have been superior to any flowerless plants of the present day.

The cone-bearing trees, somewhat like our modern pines, were the only known specimens of the upper division of flowering plants. Pine-trunks have been found in coal-measures, over forty feet in height.

No palms were yet created, so far as we know. It was once thought that the fossil remains of a palm had been found, but the tree was afterwards decided to have been only a kind of pine.

In animal remains, advance is plainly to be seen. Not only do we find, as before, creatures belonging to all the lower classes living in the sea; not only insects living on the land; not only fishes inhabiting the ocean; but also another upward step.

There is a kind of half-class, midway between Fishes and Reptiles. A crocodile and a snake are True Reptiles, cold-blooded and air-breathing. A frog and a salamander are very nearly True Reptiles. They are cold-blooded, and during part of their lives they breathe air; but during another part they breathe in water like a fish. They are therefore called Amphibians.

Many remains of large ancient lizards or salamanders* have been discovered. They seem to have been covered with hard shiny scales. One was three feet and a half long; and another had a skull seven inches wide.

A slab of rock was found in the Coal Strata of America with the footprints of such an animal dented and hardened upon it—six distinct steps of feet about four inches broad, the hind-foot four-toed, the forefoot five-toed or five-fingered. Strange remembrance this of a bygone age, when the ungainly creature walked over the soft mud of the half-built American continent, leaving his 'footprints in the sands of time' for future generations of men to see and consider.

Another variety of this animal † seems to have had webbed feet and to have lived entirely in the water. This is the earliest known specimen believed to have belonged to the True Reptiles.

We are now reaching the close of the great Ancient-Animal Period—the Period of much ocean

---

* Labyrinthodonts. † Enaliosaur or Sea-Saurian.

and little land, the Period of Lower Creatures and Fishes, the Period of Non-flowering Plants, the Period of a wide-spread mild climate, the Period of Rock and Coal preparation, the period of lower-story building in the house which was to become the Home of Man.

Much difficulty has been found in Great Britain in drawing the dividing-line between the Primary Rocks and the Secondary Rocks.

At one time the last of the Primary and the first of the Secondary were classed together as the New Red Sandstone. Both these have in them a large amount of red sandstone, and are alike scantily furnished with animal remains. It was, however, found on examination that such animal remains as did exist in the lower rocks were much more closely related to Primary than to Secondary animals. So the dividing-line was drawn above them.

Some faint signs have been seen, in the closing rocks of this first great period, which *may* point to a change of climate and to floating ice in the neighbourhood of England.

The end of the Ancient-Animal Period is marked in America by signs of grand disturbances. In the course of these disturbances the Alleghany Mountains were uplifted, and the vast coal-beds were greatly heaved and tilted and 'faulted.' Enormous pressure must have been used so to bend and curve, to crumple and fold them. 'Faults' on a tremendous scale may be seen, as, for instance, in one place, where a slip in the rock-layers has taken place, twenty miles long and twenty thousand feet deep ; yet the crack is so narrow that a man may stand astride it, with one foot on each side.

Some hold that all these movements were probably slow and gradual, but the theory may well be doubted in the face of such landslips as these. We have however, no means of coming to any certain conclusion.

Thus in some parts of the world by great upheavals and downsinkings, in other parts of the world by quietly-continued rock-building, in all parts of the world by the dying out of old kinds of animals and the entrance of new kinds upon the scene, the Primary Period passed away, and the Secondary Period was ushered in.

CHAPTER XVI.

THE AGE OF REPTILES.

'Him who alone doeth great wonders.'—PSA. cxxxvi. 4.

LEAVING behind us the great 'Ancient-Life' Period, or, as we may call it, the Ancient History of Geology, we reach the great 'Middle-Life' Period, or, as we may call it, the Mediæval History of Geology. And a strange time this Mediæval Period was upon the earth!

It is no fairy-tale which I have to relate; but the wildest fairy-tale of heroes and dragons, griffins and monsters, never surpassed the wilder reality of the earth's inhabitants in those days. Only the human hero was lacking. The dragons were there, but no St. George. He would have found foes enough to test his prowess in the unfinished land of future Merrie England—strange uncouth fantastic monsters, enormous in size and terrible to look upon, with shiny scales and big eyes and great teeth ; some with long snake-like necks ; and some with fins for use in the water ; and some with legs for use on land ; and some with wings for use in the air.

Yes, actually wings! Not birds were these creatures but huge winged reptiles—the dragons of fairy-tales existing in real life. I am conjuring up no fancy legend. The fossil bones of these enormous brutes are found strewn through the rocks of that period, and their footprints are to be seen by thousands in the hardened mud. No mistake about the matter thus far is possible, though of course many mistakes are more than possible about the precise form of any particular animal. But that such creatures did live we know—immense sea-serpents in the ocean, tremendous crocodiles in the rivers, gigantic bird-reptiles on the land, fearful flying dragons through the air.

The biggest lizard ever now seen in the hottest country in the world is not more than three or four feet in length, and the longest reptile of any kind is little over thirty feet, while by far the greater number are under ten feet. In our quiet old England no wild beasts prowl—not even a wolf—to frighten a village child in the loneliest place.

Fish-reptiles and bird-reptiles are creatures unknown, and flying dragons are laughed at as only the creation of some excited brain.

But in those days British rivers were haunted by crocodiles forty feet and more in length, with huge swimming lizards to match. Bat-like dragon-like winged reptiles of all sizes, from a few inches to over sixteen feet in stretch of wings, flapped through British air. And over British land stalked enormous bird-reptiles, having four legs, but often rearing themselves up, gorilla-like, to walk upon two. Verily a fearful place to live in would Britain have been then, and not Britain only, for the same gigantic monsters ranged over the world, through the European and American continents alike. We may be thankful that they died out before the creation of man.

One cannot help wondering whether, although they did so die out in very great measure, by or at the conclusion of the Middle Period, some few kinds may not have long lingered, becoming more and more rare. Crocodiles and the smaller sorts of lizards do last up to the present day. It is not absolutely impossible that a small number of the strange and uncouth scaly monsters, flying through the air and striding over the land, *may* have remained through succeeding ages, each generation smaller than the last, yet not all completely dying out until after the creation of Adam.

If only one or two such animals had been seen by only a few men in earliest times of human history, it would have been quite enough to give rise to all the mysterious legends and wild nursery tales of dragons and griffins, handed down through uncounted generations, and coming from nobody knows where.

This is only conjecture, and we have no proof that so it was. But the fact that no remains of such creatures have been found in rocks of later date, is no absolute proof to the contrary. It only proves that if such creatures *did* survive through later ages—which they may or may not have done—their remains were either not preserved at all, which is quite possible, or geologists have not examined those rocks in which they lie, which is quite possible also.

Some of the animals which haunted the earth in Secondary Days were Amphibians, though the greater proportion were True Reptiles. The Amphibians seem to have come to their climax, as to size and numbers, in the earlier part of the Period, and afterwards to have gradually lessened.

The hand-shaped tracks of one kind* are plentiful in America. Many of them walked on all-fours, as lizards do now, but some seem to have gone on their

---

* Labyrinthodonts.

hind-legs, rarely bringing their fore-feet to the ground. One of these big fellows, marching on his hind-feet over American mud, has left eleven distinct foot-marks, each one no less than twenty inches long, while the length of his stride was as much as three feet.

Our little smooth-skinned lizards and salamanders, of modern days, would have reason to tremble at the sight of their enormous upright scaly relative of olden times. Tiny lizard-like animals, however, existed then also, for their tracks are found, not over a quarter of an inch across.

Another of the larger specimens found in Europe had a skull two feet in length, and teeth three inches long.

Both Europe and America abounded in Swimming Saurians. By a 'saurian' is meant simply a 'scale-covered reptile.'

Some of these ancient water-lizards were of immense size. The remains of one kind* show it to have been often from ten to forty feet in length, with paddles much like those of a tortoise or a whale, very big eyes and powerful teeth, sometimes as many as two hundred in number. The remains of fishes and reptiles are found inside the skeletons. Another, somewhat similar, had an enormously long neck, with

* Ichthyosaur.

a small head at the end of it, teeth like those of a crocodile, and large paddles. The skeleton of this creature,* too, has been found in Europe up to thirty or forty feet in length.

Many huge swimming-lizards lived in the ocean, and to them has been given the name of Sea-Saurians. Another inhabitant of the ocean was a long snake-like reptile,† covered with bony plates, and having small paddles. Several fossil sea-snakes of this description have been discovered in American rocks, once overflowed by the sea, and the largest of them was nearly eighty feet long.

Besides these and other great water-animals, there were crocodiles in rivers and on land, more like crocodiles of the present, but far surpassing them in size. A certain American specimen is said to have been at least fifty feet long, and to have stood ten feet high, a monster hardly outdone, one would imagine, by any other monster of even that extraordinary age.

The fossil bones of another singular creature‡ have been repeatedly found, and its foot-marks are seen by hundreds. This animal is described as having been 'crocodile-like' and also 'bird-like.' Sometimes it was thirty or forty feet in length. It had four legs, but the hind-legs were much stouter and stronger than

---

* Plesiosaur. † Mosasaur. ‡ Dinosaur.

the fore-legs. It seems often to have reared itself up, kangaroo-like, and walked on the hind-legs alone—truly a fearful sight to look upon. The fore-feet had four toes, but the hind-feet only three; so at first when these hind-foot tracks were found alone they were taken for those of a huge ostrich-like bird; and, indeed, when the skeleton was first found, the bones also were supposed to belong to a bird, from their hollow make. Some think that though a reptile it may have been warm-blooded. This gigantic ostrich-kangaroo-lizard—how else describe it?—inhabited both Continents, including the island of Britain, but it perhaps abounded most in America.

In addition to the wingless bird-reptiles, we learn from the rocks and their buried fossils of Flying Saurians,* or Dragons as they may truly be called. The fossil remains of one show it to have been three feet across the outspread wings, having a body one foot long, with hollow bones like those of a bird, skin, claws, and teeth like those of a reptile, and wings like those of a bat. This particular animal was, however, but a small specimen. Remains of such dragon-like creatures have been discovered in Europe measuring sixteen feet, and in America measuring twenty to twenty-five feet, from tip to tip of the outstretched wings.

* Pterodactyls or Pterosaurs.

An American turtle of those same days was no less than fifteen feet broad, between the tips of the flippers. Yet the naturalist who examined the fossil saw signs of its having been young at the time of its death. What must the full-grown turtle have been, if this were indeed only a gigantic infant of its kind?

Thus in Secondary times the Reptiles had chief possession of all three elements. They were the largest and the most powerful, alike on land, in water, and in air.

In the Connecticut Valley of the United States, on some sandstone rocks belonging to this period, tracks of animals are found by thousands, footprints of lizards big and little, footprints of reptiles, birds and other creatures, in different places, through some eighty miles of country, and repeated downwards through some eighty feet of strata. This tells of a long period, during which the animals lived and the rocks were built. Probably it was a low level of country, sometimes consisting of dry mud, sometimes covered with shallow water.

Such footprints are also found in Europe, though in smaller numbers.

The Middle-Animal Period is commonly divided into three lesser Ages.

In the earlier rocks there is a great deal of sandstone and limestone, and also there are a few coal-strata.

The Great Coal Age was over, but coal preparation was not entirely at an end, if indeed it ever has ended, from that time to this. Only it was no longer the *leading fact* in the history of earth. The large cellars of the house, if we may so say, were nearly full, although from time to time another cart-load was still added to the store.

About the middle part of the Secondary Period, there seems to have been coral-reef-building in or rather over England, showing a warm sea.

In addition to the abounding reptile-life of those days, there were plenty of sea-creatures and plenty of fishes. The fishes in general were of a higher grade than most of those in Primary Days.

The rounded wheel-like fossil of the Ammonite is perhaps better known than any other kind of fossil It was in this Middle Period that the Ammonite flourished, much as the Trilobite had flourished in earlier times.

All three-toed fossil footprints were at first taken for those of birds. It is now believed that many belonged to reptiles. Still we have full proof that birds did live in those days, for their stony remains have been discovered.

At a certain place in Bavaria, for instance, a great

many fossils were found, belonging to the Secondary Period. Among flying reptiles, tortoises, fishes, insects, and others, appeared one very perfect bird-skeleton. So well had it been preserved, that some of the feathers were there with the bones.

We thus have at last all the lower classes of the highest sub-kingdom. There are Fishes, Reptiles, and Birds. Step by step we have followed upward the order of Creation, as it seems to have taken place. Only Mammals remain wanting.

And wonderful to say, though the Age of Mammals was yet to come, the first mammals actually ived in the Age of Reptiles. Both in Europe and in America fossil-remains have been repeatedly found of a small insect-eating quadruped of this class. It is believed to have been an animal carrying its young in a pouch, just as the kangaroo does. However this may have been, the fact that it belonged to the highest class —that of MAMMALS—seems clear.

The words 'Mammal' and 'Quadruped' have not exactly the same meaning. Every quadruped or four-footed animal is not a mammal, and every mammal is not a quadruped. Crocodiles have four feet, but they are not mammals. Whales and seals and human beings have not four feet, yet they are all included in the Mammal Class.

Here again, as in the case of Fishes, we have the sudden appearance of a perfect animal, higher in rank than any earlier created, in the great Period just before that in which its class becomes the chief class of all.

But though the little Quadruped was of highest rank among beasts, its weakness and smallness were remarkable, in contrast with the gigantic strength and size of its reptile companions.

Plants did not vary greatly through the greater part of the Middle-Life Period from those of earlier days. There were ferns and horse-tails and pines still, with the addition of a kind of cypress. No true grasses, or mosses, or superior plants have yet been discovered. We shall, however, find a marked and sudden change towards the close of the Period.

CHAPTER XVII.

THE AGE OF CHALK.

'The Lord which . . layeth the foundation of the earth.'—Zech. xii. 1.

THE third and closing division of the Middle-Animal Period is commonly known as the Chalk Age. The Latin for chalk is 'creta,' and these rocks are called 'Cretaceous,' because a large portion of them consist of chalk.

This was the great Chalk-making Age in the world's history, and the amount formed in that one age was something enormous. The white cliffs running along the southern coasts of England are a very small part of the whole chalk formation. It extends from the north of Ireland to the Crimea, a distance in one direction of eleven hundred and forty miles, and from the south of Sweden to the south of France, a distance in another direction of eight hundred and forty miles ; while in places it is hundreds of feet thick.

When we remember that the whole of this vast mass is made up of tiny white shells, each one formerly the home of a minute living animal, the amount is amazing.

It is rather a sudden change to turn from gigantic crocodiles and winged monsters to these smallest of creatures—though we have not really done with the monster reptiles yet, since they lived and flourished all through the Chalk Age. But undoubtedly *the* great Fact in that Age was, as the name implies, not the existence of huge reptiles with paddles or wings, but the life and death of countless millions of almost invisible specks of life.

It is remarkable to find this backward step, if we may so term it, in the stately onward march of events. Hitherto we have had to follow a regular stepping upward, a progression stage by stage from lower to higher forms, not merely in the order of actual creation, but in the order of the principal or dominant kinds of life upon earth.

First, Boneless Lower Animals ; then Fishes ; then Amphibians ; then Reptiles. Next would come Mammals ; but before the Age of Mammals, and breaking in, as it were, upon the Age of Reptiles, we have the Age of Chalk—in other words, a later Age of lower animals. Enormous past imagination must have been the numbers of these tiny creatures, swarming in the ocean, and doing their work,—not much work either, since each one had only to live and then to die, when the shell sank to the bottom amid myriads of others, to become part of the uppermost layer forming there. But who shall venture to say that the mighty brutes of land and sea were playing a more important part in earth's history than these small animalcules?

Although the Chalk formation reaches so many hundreds of miles, north and south, east and west, through Europe, it is not always visible. Here it rises to the surface of the ground, showing as high white cliffs; there it sinks below later-built rocks, and is hidden from sight; again it rises and becomes visible, and then again it disappears.

For a long while there was uncertainty as to the exact manner in which chalk had been made. A little of the fine white dust of chalk, rubbed off and put under a powerful microscope, was found to be chiefly composed of tiny shells, some whole, some broken. The greater number of these were rhizopod-shells.[*] It was supposed that chalk was formed

[*] Foraminifers.

in deep water, of shells quietly dropping and collecting; not, like the harder kinds of limestone, made in shallow water, of shells ground up into fine dust by the waves. Chalk is a kind of limestone, but it is of a soft loose make, not hard and compact.

This guess or theory was proved to be true, when deep soundings were made in the Atlantic Ocean, before the Electric Cable was laid down between Ireland and Newfoundland. Mud, dredged up from the bottom, at a depth of two miles, was found to be chiefly composed of tiny shells, bound together by a kind of living gelatine, and thus becoming gradually real chalk. Nineteen-twentieths of them were rhizopod-shells.

But there is another puzzle about these white chalk cliffs. If you examine them carefully, you will often see, not far from the top, a line of flints, sometimes placed close together, sometimes scattered loosely along at the same level. Then there comes a band of chalk, and then again, perhaps, another layer of flints, and this may be several times repeated. And if you were to break open a great many of these flints, you would find some of them to contain the fossils of living creatures.

But how could living animals by any possibility have crept inside the hard flints?

The question comes to this—How are flints formed? What are flints made of? Did the animals get into the flints while alive, or did the flints form round the animals after they were dead? The latter certainly appears the more likely of the two.

The mere presence of these layers in the cliffs is of itself a perplexing matter. There seems no connection between the flint of the layers and the lime of which the chalk cliffs are made.

We may not yet speak positively on the subject, but the deep soundings and dredgings of the Atlantic have furnished a possible clue to this riddle also.

The dredged-up mud was found, as just stated, to consist chiefly of foraminifers, or the shells of one kind of rhizopods—chiefly, but not entirely. For while in certain parts the Atlantic bottom seemed to be quite in the possession of the rhizopods, in other parts they were nearly absent, and the tiny vegetable diatoms* had the field to themselves.

The diatoms, although believed to be plants, have their minute shells, just as most rhizopods have. But while the foraminifer-shells are made of lime, the diatoms' shells are made of flint. The following has been offered as a possible explanation. At the bottom of a tolerably deep sea, where rhizopods were

---

* See p. 49.

living and dying in countless myriads, multitudes of their shells would be ever sinking and helping to build up a floor of white chalk. Animals of larger size would also live and die in the waters, and many of their stony remains are found in the chalk layers. Then there would come a change—how brought about we do not know—and where the rhizopods had abounded the diatoms would next for a while abound, their flinty shells taking a turn at sea-bottom building. They would not flourish in sufficient quantities to make deep strata, like the rhizopods, but as they sank to the ocean-floor, they would collect round shells or sponges lying there, or gather into small masses, and thus would the hard flints be formed, later to be covered by fresh layers of chalk. The tendency to collect round some small hard substance, and to become united to it, has often been observed, alike in minute shells and in fine sand or earth.

There is, however, a difficulty in admitting this explanation as very probable or conclusive, for the structure of flint, examined under a microscope, does not show a collection of flinty shells, but a *solution* of flint—that is, flint which has been dissolved,— often containing the tiny spores of non-flowering plants.

If chalk and flint are now being formed below

the Atlantic, a question may present itself to some minds,—Are we then still in the Chalk Age?

No, certainly not. The Chalk Age ended long ago. We have already seen that Coal-making took place in the Secondary Period, yet that was not the Age of Coal. Each separate Age had its great leading Facts; its principal Rock-formations, and its chief kinds of Animal life; but other rock-building went on, and other kinds of animals lived and flourished, alongside with them, though in a lesser degree; and so it is now.

Foraminifers are thus the chief material of which chalk is composed. It has been calculated that a cubic inch of chalk contains often about one million shells, yet even they are much larger than the diatom-shells. What must be the countless multitudes in the thousands of miles of the vast Chalk-formation!

How long a time was occupied by the building up of the whole, it would be vain to attempt to calculate. Such calculations sink into mere loose guesses, built upon an 'if' which is built upon nothing. We do not know with any certainty how quickly chalk may in the present be formed; and if we did, that would not help us with regard to the past. In the great Chalk Age the ocean probably abounded with rhizopods to an amount never equalled at any other time; and the work of building up would have been in that case much more rapid than it seems to be now.

In America, although there is the so-called Chalk-formation, it consists chiefly of green-sand, clay, limestone, and other rocks, but not of pure white chalk. In Europe, although sandstone and limestone are found, chalk is the principal material.

The geography of the Age has points of interest The Chalk of Europe shows that broad tracts of land still lay under water—deeper perhaps in parts than in previous times. Europe could indeed have been scarcely a continent then, in the true sense of the word, but was rather an immense collection of islands, with its largest amount of dry land to the north. The Alps and the Pyrenees in Europe, the Himalayas in India, the Rocky Mountains and the Andes in America, were still all beneath the ocean or only raised a little way out of it, possibly the higher peaks showing, but no more. A great part of the south of England and the north of France had salt water rolling over them, filled with millions upon millions of rhizopods.

With regard to Vegetable-life a change comes with the Chalk Age, and trees and plants of modern days burst upon us in full-blown perfection, with really startling suddenness. Up to that date no remains have been discovered of any plants belonging to the higher classes, unless, perhaps, one or two doubtful specimens. Now, however, the monotonous repetition of only flowerless kinds suddenly ceases.

For at Aix-la-Chapelle, in rock-layers belonging to nearly the close of the Chalk Age, a large quantity of plant-remains have been found. Among them are as usual the fossils of ferns and pines. But in addition there are fossils of the Oak, the Beech, the Sycamore, the Poplar, the Willow, the Walnut, the Myrtle, the Fig, the Maple, the Magnolia, the Holly, the Hickory, the Banksia, and others. Also the first fossil palm-leaves have been met with on Vancouver's Island, in rocks believed to belong to the same period.

With regard to Animal-life, the great reptiles continued to abound, side by side with the tiny rhizopods. Snake-reptiles and swimming saurians in the ocean, crocodiles of old and new kinds in the rivers, flesh-eating and plant-eating bird-reptiles on the land, flying reptiles in the air,—all these lived still, as through earlier Secondary ages, but they were nearing their destruction. For a while they flourished side by side with the fig, and the myrtle, and the magnolia. Then, though the frail plants lived on, the mighty beasts came to their end.

Birds of large size seem to have existed in considerable numbers. Several fossil bird-skeletons have been found in America; one of a diver, five feet and a half high.

Among the living creatures which ranged the ocean, none is more interesting than the beautiful spiral ammonite. The first ammonites, so far as we know, were in the Age of Fishes, but they were then rare. It was not until the Secondary Period that they abounded. One fossil ammonite, found in English rocks, is no less than a yard in diameter; and from this the fossils range downwards to tiny specimens an inch or less across. Sometimes they are found uncoiled, but more commonly coiled.

The ammonite, like the trilobite of earlier times, has died completely out of existence, and is no longer to be seen alive. A few specimens lingered on after the Secondary Period, but the Age of Ammonites ended with the Age of Chalk.

At the close of the Chalk Age, which is also the close of the Secondary Period, a great and remarkable destruction of life seems to have taken place.

Whether this destruction was sudden or gradual is a question about which opinions differ—the simple fact being that nobody knows.

Some believe that thousands of animals were swept out of existence in a short space of time; since they are found to have lived up to the close of the Chalk Age, and then—they are no more seen.

Others suggest that the seemingly abrupt change is caused only by intermediate rock-layers having been washed away. They argue that animals probably lived and died through long ages between—the record of which has been lost—one kind after another slowly disappearing and giving place to new kinds.

All we actually know is that such a break—abrupt in appearance, at all events—does exist. We have the Chalk Rocks, the last of the Middle-Life Period, full of animals of Secondary times. Immediately above them we have the more modern rocks, full of animals of 'New-Life' days, all entirely different from animals of Middle-Life days.

It is easy to imagine possible rocks between the two, belonging to a possible intervening Age, filled with possible animals, partly like those in the uppermost of the Middle-Period Rocks, and partly like those in the lowermost of the New-Period Rocks, thus forming a link to unite the two, or a bridge to span the chasm. But imaginations are not facts to build upon, and as yet such half-way rocks have not come to light.

It is true discovery has been made of certain layers which seem to date later than the Secondary Chalk, and earlier than the Third-Period sands and clays.

Each of these, however, was found to contain distinctly either the animals of Middle-time, or the animals of New-time, and *not* the animals of a stage lying between the two. So the layers were clearly only a part either of Middle-Life or of New-Life Rocks, and the position of affairs remains just the same as before. The chasm stands as wide as ever. We still pass at a leap from one Period to the next.

In all Europe, in all Asia, and in nearly all America, not one single kind of living creatures in Third-Period Days has been found, precisely like one single kind in preceding Secondary Days. In the Rocky Mountains only, some specimens may be the same, though even this is doubtful.

It is not, of course, absolutely impossible that throughout this vast extent of country great masses of rock, lying between the two Periods, *may* have been washed away—a whole chapter, as it were, neatly torn out of the book. But though not impossible, it would certainly be very extraordinary. Such a tremendous and wholesale destruction of strata, over Europe, Asia, and America, at one particular point of time in geological history, would be at least as remarkable, in whatever manner it came about, as the most wholesale and sudden destruction of Animal-life through the world at that same point of time.

Which of these wonderful events, however, really took place we cannot say. We only see the gap—the break—the change. We only see that animal-life did so pass through a transition—some kinds dying quite out—some kinds remaining on, the same yet not quite the same; and with them the creation of entirely new kinds, higher in the scale than any before.

Signs are sometimes observed in the chalk of possible ice-action. Certain big stones and boulders, clustered here and there amid chalk-layers, are exceedingly perplexing, except under the supposition that they were dropped upon the half-formed chalk by icebergs floating as far south as English seas.

This, if true, seems to tell of a possible change of climate in the earth, towards or at the close of Secondary days. It *may* be that such a change, from long-continued mildness to cold, was the cause of the wide-spread destruction of life.

Here again the question arises whether such a change of climate would have been sudden or gradual, and here again we cannot reply with any certainty. A gradual change would mean only a gradual passing away of one kind after another. A sudden change would mean sudden and wide-spread death. It is a singular fact that signs of such a death, yet of a death which has not injured the bones, has been noted by geologists in the reptiles of those days; just as it was noted in the fishes of earlier times.

The upheaving of mountain-ranges, the rising or sinking of the ocean-bottom here or there, might have entirely changed the directions of warm or cold ocean-currents, and might thus at any time have brought about a complete alteration of climate in any land or continent in the world. Such risings or sinkings again might be either sudden or slow. But beyond broad statements of what 'might have been,' particular guesses and closely worked-out theories are worth little.

This chapter can scarcely be better ended than with the following beautiful lines, descriptive of the Middle-Life Period:

> 'The Nautilus and the Ammonite
>   Were launched in storm and strife;
> Each sent to float, in its tiny boat,
>   On the wide wild sea of life.
>
> 'And each could swim on the ocean's brim,
>   And anon its sails could furl;
> And sink to sleep in the great sea deep,
>   In a palace all of pearl.
>
> 'And theirs was a bliss more fair than this
>   That we feel in our colder time,
> For they were rife in a tropic life,
>   In a brighter happier clime.

'They swam 'mid isles whose summer smiles
    No wintry winds annoy ;
Whose groves were palm* (?) whose air was balm,
    Where life was only joy.

'They roamed all day through creek and bay,
    And traversed the ocean deep ;
And at night they sank on a coral bank,
    In its fairy bowers to sleep.

'And the monsters vast of ages past,
    They beheld in their ocean caves ;
And saw them ride in their power and pride,
    And sink in their billowy graves.

'Thus hand in hand, from strand to strand
    They sailed in mirth and glee,
Those fairy shells with their crystal cells,
    Twin creatures of the sea.

'But they came at last to a sea long past,
    And as they reached its shore,
The Almighty's breath spake out in death,
    And the Ammonite lived no more.

'And the Nautilus now, in its shelly prow,
    As o'er the deep it strays,
Still seems to seek in bay and creek
    Its companion of other days.

'And thus do we, in life's stormy sea,
    As we roam from shore to shore,
While tempest-tost, seek the loved—the lost—
    But find them on earth no more.'
                          G. F. RICHARDSON.

---

* It was once believed that palms existed much earlier than is now supposed.

CHAPTER XVIII.

THE AGE OF MAMMALS.

'Which by His strength setteth fast the mountains, being girded with power.'—Psa. lxv. 6.

WE have now reached the last great division of Geological History.

It will be remembered that the whole of this strange Fossil-History, written in the rocks, is divided into three portions.

Sometimes the three are described as Primary, Secondary, and Tertiary; the meaning of the terms being simply, First, Second, and Third.

Sometimes they are described in words which when translated, mean, Ancient-Life Period, Middle-Life Period, and New-Life Period.

Again, they may be described as the Period of Lower Animals and Fishes, the Period of Reptiles, and the Period of Mammals.

Or, once more, we might term them the Ancient

History, the Mediæval History, and the Modern History, of Geology.

It is upon the last of these three that we are now entering.

Although the Tertiary or Third Period lies nearer to our own days than any before, yet the animals which lived throughout the greater part of it were very different from those which live now. Not one single fish, or reptile, or bird, or quadruped, known to have existed then, was exactly the same as any one fish, reptile, bird, or quadruped, existing now.

But this was not quite the case with sea-shells and fresh-water shells. At the beginning of the Tertiary Period, indeed, very few even of these were the same as any seen now; still there were a few, and the number went on gradually increasing. By the close of the Tertiary, and the beginning of the Post-Tertiary or After-Tertiary, about ninety-five in every hundred shells were exactly the same as those of the present time.

So the first part of this Third Period is called the 'Dawn of Recent,' and the next has a name which means that 'Less' than half of the shells were 'recent,' and the third has a name which means that 'More' than half were 'recent,' and the last name means that 'most' were 'recent.'

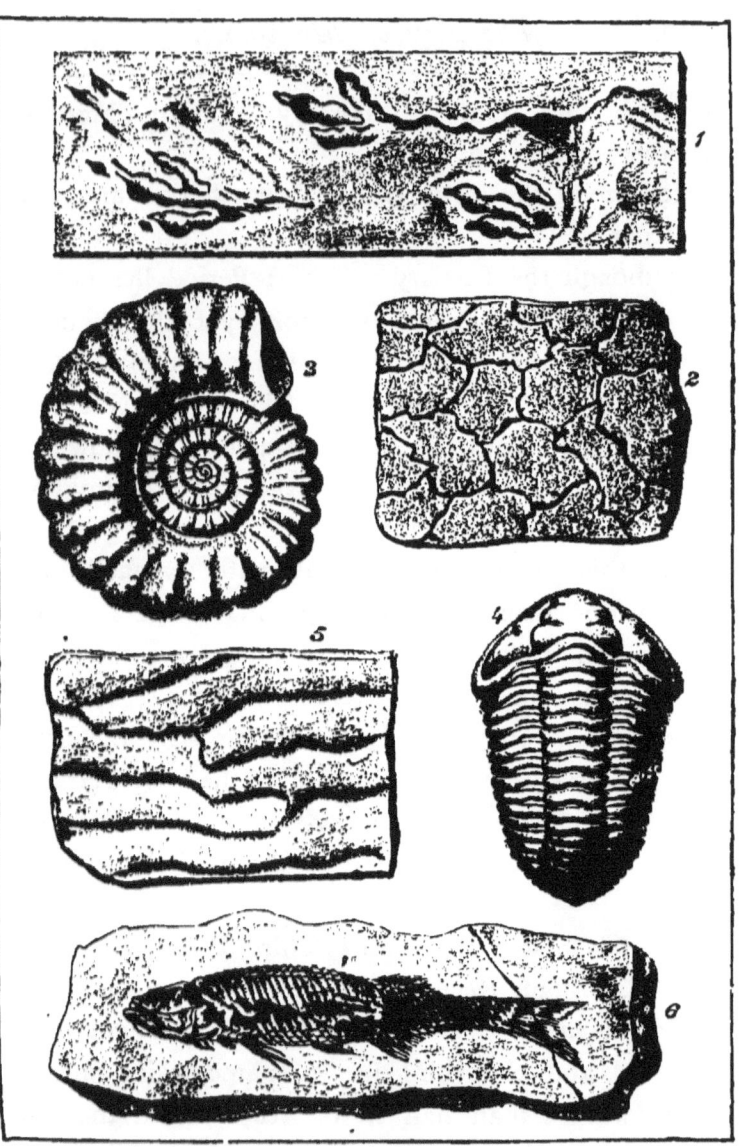

FOSSILS.

1. Footprints of a Bird.   2. Mudcracks.   3. Ammonite.
4. Trilobite.   5. Ripplemarks.   6. Fish.

The rocks of this Period in Europe and in America bear a general likeness one to another. They lie close over the Chalk Formation, with the marked break between the two, as to kinds of animal remains, described in the last chapter.

An exception is found to this widespread 'break' in the Rocky Mountains, where the change from Secondary rocks to those above is gradual.

Fossils abound in the Third-Period rocks, though more in some parts than in others. There are layers of earth entirely made up of animalcule-remains, so small as to look like fine dust to the naked eye. Also rocks, to a vast amount, are largely composed of a kind of rhizopods, not tiny in size like those described in earlier chapters, but flat, coin-shaped, and about as large as a threepenny or sixpenny piece. They are called *Nummulites*, from two words, one Latin and the other Greek, meaning ' coin ' and ' stone.'

The wide reaches of country over which nummulites are found, show that these small creatures must have lived in multitudes past imagination. The Day of Rhizopods, which came in during the Reptile Age, seems to have continued through part of the Mammal Period.

Some limestones consist almost wholly of nummulites. Rocks containing these coin-shaped fossils extend throughout a great part of south Europe, south Asia, and north Africa. They are found in the Alps, the Pyrenees, the Apennines, the Carpathians, the Himalayas, also in Egypt, in Afghanistan, in Persia, in Thibet, in Japan, in Java.

Moreover, though the animals must all have lived and died in ocean-waters, they are found at great heights above the sea-level. The summit of the Dent du Midi, for instance, over ten thousand feet high, is formed of nummulite rock.

If then, as is believed, all the rocks holding these particular fossils, were formed in the course of the New-Life Period, it follows that the Alps, and the Pyrenees, and parts of the other ranges just mentioned, were not mountains at all until after the beginning or middle of that Period. And this is supposed to have been really the case. The kingly Mont Blanc and Monte Rosa were probably buried under ocean-waters, and not uplifted as lofty mountains until towards the close of the Third Period.

But how long were the ages which may have been occupied, first by the building of the coin-filled rock layers, and then by the grand upheaval of whole mountain-ranges, we cannot even guess. We do not know in the least how quickly the former may have taken place, or whether the latter was a gradual or a rapid work.

It should be understood that these stone coins and other sea-fossils are not merely found scattered over the surfaces of great heights, but are buried deep down in the solid rocks.

The climate of the earth in Third-Period Days seems generally to have been mild, much milder than now. Tropical plants grew on European earth, and tropical animals wandered through European forests; while plants which now flourish in temperate lands then flourished within the Arctic circle.

Towards the close of the Tertiary a change becomes visible; showing a gradual increase of cold in the hitherto warm waters and soft atmosphere of England and middle Europe. The Third Period seems to have been possibly ushered in by a cold era; and it seems to have almost certainly gone out in a cold era. But the chief part of it, not counting the beginning or the ending, appears to have been warm.

As in other ages, there was most likely much gradual building of sediment beds, sandy, or clayey, or muddy, in Europe and America, with frequent rising and sinking of land in different parts. The continents would by this time have had their general outlines much the same as at present, though broad reaches which are now dry land must then have lain under the sea for at least a part of the Period.

Rocks of this Period are to be seen in the neighbourhood of London and Paris, among many places. The 'London Basin' and the 'Paris Basin' are terms often used; the 'Basin' being in fact a large hollow in rocks of the Chalk Age, which hollow has been filled up with later-built layers of Third-Period sand, clay, gravel and other materials, containing ample fossil remains. The 'Paris Basin' is about one hundred and eighty miles long and ninety miles broad.

The British Isles, like other lands, are believed to have slowly risen and sunk alternately through many ages. At one time the whole was probably part of the Continent; then again it sank, and became a mere little archipelego of small islands; again it rose, and Ireland was separated from England, and the Dover Straits flowed between England and France. In early New-Life ages England was probably part of the Continent, since all the great European beasts found their way across into Britain; but the final separation probably took place towards the close of the Period.

CHAPTER XIX.

MORE ABOUT THE AGE OF MAMMALS.

'Thou sendest forth Thy Spirit, they are created; and Thou renewest the face of the earth.'—PSA. civ. 30.

SOME few signs are seen in early Third-Period rocks, which may point to the existence of icebergs floating southward, but the general evidence throughout the greater part of the Period seems to tell of a mild climate.

Towards the close, however, a marked change becomes visible. Rough 'drift' or 'till,' polished stones, scratched rocks, big blocks of stone carried far from their places, coarse clay,—these and other tokens found in the upper layers speak to us of a colder time coming on.

Just as the First Period was the Age of Lower Animals and Fishes; as the Second Period was the

Age of Reptiles; so the Third Period was the Age of Mammals.

Mammals have been already described as the very highest class in the upper division of the Animal Kingdom. They have a back-bone and skeleton; so have fishes. They breathe air; so do reptiles. They are warm-blooded; so are birds. They suckle their young; and so do not any other living creatures. This is the mark of uppermost rank. The mammals, or the animals of highest rank, were the latest created.

The Tertiary or Third Period is sometimes spoken of as ending before the Post-Tertiary or After-Tertiary, and sometimes it is made to include the Post-Tertiary. In the former case it is the Age of Mammals, because mammals first flourished during its three ages. In the latter case it is still more strongly the Age of Mammals, for in the Post-Tertiary these animals reached their greatest size and power. Moreover, Man was created in the Post-Tertiary, and Man is a Mammal.

In this chapter, however, we are thinking about the Third Period, apart from the After-Third Period.

During the earlier part of the Third Period ammonites, as well as coin-like nummulites, lived over the submerged Alpine peaks—but they were the last of their kind. Coral-polyps too worked there, showing a warm sea.

The fig-tree and the cinnamon grew wild in the Isle of Wight; and palm-trees, of a kind now seen in Bengal, grew in the Island of Sheppey; and the cocoa-nut and the custard-apple grew near London.

These facts are known by the fossil leaves and fruits or seeds found in the rocks of the different places. It would hardly be believed, without being seen, how perfect and delicate is the way in which some of these fossils are kept. A fossil butterfly was found in Croatia, the very pattern of the wing being exactly depicted on the stone. Thus it is that botanists can tell the trees or plants to which fossil leaves and stems and seeds probably belonged.

In England there were crocodiles, and sea-snakes up to twenty feet in length, besides tortoises, turtles, and other animals. Also a bird, found in the London clay, had notches in its bill resembling teeth. Nor is this the only specimen which has been discovered, of a 'toothed bird' of former days.

In the forests of France there were parrots and flamingoes, cranes and adjutants, pelicans and secretary-birds—such as are now brought from Africa and other hot countries to put into Zoological Gardens. A kind of ostrich too seems to have lived there.

In the Rocky Mountain region where the break between Second-Period and Third-Period Rocks is not seen, but where one age seems to have glided gently into the next, the marked change visible elsewhere in kinds of animal-life does not exist. Chalk-Age animals, chiefly reptiles and not mammals, continue; so it is really very difficult to say where the rocks of the Chalk-Age end and the rocks of the Recent-Life Period begin.

But in other parts of North America and in Europe, alongside with the birds and the few reptiles above-mentioned, mammals flourished in great abundance. Many of them were strange creatures, utterly unlike any that live now. Others resembled more or less modern quadrupeds, though they were never exactly the same.

One of these brutes, the remains of which are often found, has a name given to it which means 'Ancient Wild-beast.'* It was something like the modern long-nosed tapir; the larger specimens being as large as a horse, and the smaller as small as a sheep. There were many different kinds of tapir-like animals in Europe then.

Another strange creature† was rather slender and graceful in form, and about the size of a chamois.

* Paleothere. † Xiphodon.

There were dogs, weasels, opossums, and numerous other quadrupeds in Europe; some of them known to us by their skeletons, and some only by their footprints. So numerous, indeed, are the kinds of footprints, that we may learn from them a useful lesson, as to the variety of animals which may have lived in any age, quite unknown to us by fossil remains.

A large whale-like animal* seems to have been abundant in America—that is to say, in the seas overflowing what is now American land. The back-bone is from fifty to seventy feet long. In one part the remains of no less than forty of these ancient whales were found within ten miles of one another. Whales and porpoises are mammals, not fishes, since they are warm-blooded, and breathe air, and suckle their young.

The European plants and trees of the early part of this Period, seem to have been generally much the same in character as Australian trees and plants of the present day; but later on in the Period they appear rather to have resembled the modern growth of North America.

In middle Europe palms grew plentifully as far north as 50° North Latitude. There palms ceased, but their companions are found in much higher latitudes.

* Zeuglodon.

In the Valley of the Rhine there were oaks and poplars, maples and planes, and a kind of North American vine.

In North Greenland, now a region of perpetual ice and snow, the wellingtonia flourished side by side with the poplar and the willow, the oak and the beech, the plane and the walnut.

In Spitzbergen, where now only a few stunted shrubs and certain lichens find subsistence, the beech and the poplar, the plane and the lime, the alder and the hazel grew.

The uplifting of the Alpine mountains and other great ranges is believed to have taken place somewhere about the middle of the Period.

Remains are found in France and in England of huge quadrupeds, though not yet so huge as those in the Post-Tertiary Age. There were the rhinoceros, and the hippopotamus; also the mastodon, a very large and massive kind of ancient elephant; also a singular kind of elephantine-hippopotamus-tapir;* also many other different kinds of big tapir-like beasts; also stags and antelopes, giraffes and camels, monkeys and crocodiles, whales and dolphins. All these flourished on land or in water during the middle part of the Third Period in Britain and

* Dinothere.

France, as well as in other parts of Europe and in America.

One gigantic extinct tortoise of those days is described as having been not less than eighteen feet long—the animal, not the shell only—and as standing seven feet high. But though some great reptiles lived still, yet their Day was over.

In the latter part of the Period, as already said, marks are seen of a change in the climate. Semi-tropical plants in Europe gradually give place to those which tell of a colder atmosphere. Also, if the rocks containing fossil-shells are examined — for instance, those of the Crag in Norfolk—this gradual change is distinctly visible in passing from layer to layer upwards: the kinds which live in warm waters yielding, step by step, to those which live in frigid waters. The mammals, however, continue much the same as earlier.

In a certain buried forest, near Cromer, once dry land, though now only visible on the shore at very low tide, the remains of many creatures are found amid the remains of tree-trunks—such as the elephant, the hippopotamus, the rhinoceros, the deer, and smaller animals. The mastodon is not discovered there, though its bones lie elsewhere in English rocks of that date.

More about the Age of Mammals.

On the European and American continents, the record still tells us of the ancient whale,* and of a kind of rhinoceros† as big as an elephant, with huge horns. Also, there were horses, somewhat like those of modern days, only one kind was as small as a fox; and all kinds seem to have had feet with more or less distinct toes, instead of the single hoof of the modern horse. Also, there were mastodons, elephants, tigers, wolves, deer, foxes, porcupines, ant-eaters, monkeys, and many other kinds. All these ranged the earth late in the Third Period; but not one of them was exactly the same as any kind living now; and, among them all are to be found no signs of ox or cow.

* Zeuglodon.   † Dinocere.

CHAPTER XX.

THE AGE OF ICE.

'How great are His signs ! and how mighty are His wonders !'
—DAN. iv. 3.

WE have now reached comparatively modern days in this strange old-world history; and it may seem as if the task of writing about those days ought to be easier than the task of writing about more distant ages—yet, in reality, it is not so. The difficulties increase with the increased nearness of the time.

This is not hard to understand. If you are looking at some object in the far distance, its outline is simple. You may be puzzled to decide exactly what that object is, but the little you *can* say about it is easily stated. If, on the contrary, you have to describe scenery near at hand, the very fulness of details makes it more difficult to give a clear description.

The same is found in the history of men. In telling a tale of ancient days, much may seem dim and doubtful ; but the whole of your knowledge concerning the events in question, whether certain or uncertain knowledge, may be summed up in a few clear sentences. But try to write a history of the days of George III., and the very abundance of facts, the complexity of accounts, the warmth of feelings involved, will add a thousandfold to your difficulties.

So it is in Geology. The nearer we approach to present days, the greater becomes the number of facts and of theories, and the more difficult it is to put all that has to be said into one or two brief chapters.

All fossil-shells found in the rocks, belonging to the *first* half of the Post-Tertiary Age, are the same in kind as those now living ; but most of the mammals have died out, or have given place to new kinds, very much like, yet not quite like, themselves.

Throughout the *last* half of the Post-Tertiary Age, commonly known as the Recent Age, not only shells, but in a great measure the quadrupeds too, agree with those now in the world.

The Post-Tertiary Age seems to have come upon

the earth in the shape of an intensely cold period—the great Ice-Age of Geology.

I have already mentioned the signs in certain later Third-Period rocks of a change of climate —warm-water shells giving place to Arctic shells, and semi-tropical plants to those of more northern latitudes.

This increase of cold towards the close of the Third Period, seems in the beginning of the following Age to have reached a climax of wide-spread and long-continued and very severe frost over the temperate latitudes.

An account was given in the eighth chapter of the Drift, or Till, or Boulder Clay—different names for the particular soil which is believed to bear the marks of a past mighty Ice-age. The scratched stones, the scored and polished rocks, the heaped-up old moraines or what appear to be such, the huge erratic blocks scattered through different countries and seemingly brought from a distance, the tough clay filled with sharp-edged stones — all these are there described.

More than once in the course of this old-world rock-history, mention has been made of stones and blocks here or there, which *may* possibly have been dropped by floating icebergs upon the half-built strata—

though it is quite as likely that some, at least, of them were carried out to sea, entangled in the roots of floating tree-trunks, or borne on a floating island, and thus dropped. In those earlier cases the evidence of plants and animals living at the time, points almost without exception to a mild climate.

But in the great Ice-Age, of which we hear so much, stone-markings, blocks, plants and shells, all unite to tell the same tale of intense and long-continued cold.

The great Ice-Age is commonly believed to have taken place somewhere near the beginning of the Post-Tertiary Age.

Of course this does not fix it at any particular *date.* In geological history we cannot count by years and centuries, and every attempt to do so sooner or later breaks down. We are only able to count by 'Periods' and 'Ages,' each 'Period' and 'Age' being of unknown length.

Some geologists have supposed that the Ice-Age may have taken place many hundreds of thousands of years ago, while others suppose it to have been comparatively quite near to modern times. No one can tell which really was the case.

We know so much as this, that in the strata commonly called Post-Tertiary—some of the latest formed, and in fact the top-story of the great earth-crust building—there are certain signs difficult to explain. Signs, such as tough clay, filled with jumbled stones of all shapes; such as scratched and polished rock-surfaces, and scratched and polished stones; such as great jagged rock-masses on mountain-tops, brought from far-off heights. No theories of rushing water or stormy waves are sufficient to explain these appearances.

Therefore it is that the belief has sprung into being of enormous and wide-spreading glaciers. For this same work of polishing, scoring and carrying, which we count glaciers to have accomplished in the past, they are seen to be daily doing in the present.

It seems a tremendous supposition that, after the prolonged ages already described of soft and warm climates over all the earth, even within the Arctic circle, there should have come a period of such amazing cold, as to bury Canada and a great part of the United States, Scotland, Switzerland, and a great part of France and England, under enormous glaciers, branching in all directions, spreading through thousands of miles, covering all lower heights, and reaching two or three thousand feet up the sides of mountains. But on the other hand, although tremendous, it is not impossible. 'Nothing is impossible' with God. And as at present we know of no other possible cause for these strange facts and appearances, the glacier explanation is accepted as in all probability true.

Another thing that we do not know is *how* this great change in the climate of Earth came about. By 'how it came' I am not questioning the fact that it came straight from God. But it is usually His will to work by means; and what means He employed to bring about the change we cannot tell. Many guesses have been made, some very ingenious ones among them; but none yet which can be honestly and fairly accepted as entirely satisfactory.

It was rather a perplexity in connection with this cold period, that though the shells and plants and ice-marks tell of a frigid climate, yet the bones of elephants and rhinoceroses are found as at that time living in England. The elephant and rhinoceros are not usually inhabitants of very cold countries.

The difficulty has been met in more ways than one. Sometimes, it is said, these creatures, though living in warm lands, are found to wander to the neighbourhood of frost and snow, as, for instance, when a Bengal tiger was found amid the snows

SKELETON OF A MEGATHERIUM.

of the Himalayas. This very rare event, however, would hardly serve to explain the abundance of huge quadrupeds in England through those days.

Again, it has been suggested that regular summer and winter migrations may have taken place then with elephants and rhinoceroses, as now with swallows. The supposition seems a very improbable one.

Again, some of the huge quadrupeds of those times are known to have had a warm covering of hair and wool, which would have fitted them to stand severe cold. As we usually know them only by their skeletons and footprints, the instances are few in which we can learn anything about their skins. Warm woolly coats may have grown upon them, when rendered needful by increase of cold, as we now see with animals inhabiting Arctic regions.

Following after the Ice-Age, we find signs of great floods in many different places. It is not impossible that these floods were caused by the rapid melting of the mighty glaciers which so long had overspread the lands.

Here again, as before, the thought of the Flood in the days of Noah rises to mind. Whether that

Flood and these floods of which the rocks tell us were the same, or whether as some think they were separated by a wide interval of time, it is not possible to decide. We know nothing as to the date of the Geological floods. But in either case, it should ever be remembered that such tremendous rushes of water over the surface of the earth must have worked many a change in the upper rock-layers, and have added fifty-fold to the difficulty of rightly reading the latest-written chapter in the Geological volume.

During the flood-time, after the Ice-Age, there was probably much active work done, in the way of cutting out valleys, and carving deep ravines. Much, though not all, of this has been the work of running water. A stream of water may flow long over a nearly level plain and do small damage, but if rushing down a steep hill-side it will tear away earth, and wear away rocks with great rapidity, deepening its channel, and cutting its path visibly lower from year to year.

Sometimes this time of floods, or a time following soon after, is called the Terrace Period. In a watercourse, the old levels of the stream—those heights at which it ran in bygone days, before it had cut its way down to its present level—are often to be seen as terrace above terrace, or shelf above shelf, on either bank. The highest terrace or shelf is the oldest level, the next lower the next oldest, and so on.

After the floods, a time is supposed to have succeeded of quiet earth and sand deposits in tranquil waters.

The uppermost sandy and earthy layers, those which have been spread over the lands by overflowing waters in latest times of all, are commonly spoken of as 'alluvium.' This is taken from a Latin word, meaning 'an inundation.'

England, in the days following after the Third Period, was indeed a different country from England of the present.

It is true she was no longer inhabited by bird-reptiles and flying-dragons; yet her inhabitants were scarcely less startling, though perhaps less grotesque.

Enormous elephants, almost twice the size of modern ones, roamed through her forests; and huge two-horned rhinoceroses kept the elephants company; and great tusked hippopotamuses wallowed in her swamps; while gigantic lions and tigers disputed the palm with the bigger beasts;

and leopards and wolves, deer and wild horses, bears, wild cats, and countless savage hyænas, together with many lesser quadrupeds, roved in her woods, skulked in her caves, fought, fled, and devoured one another. Their bones are found in abundance in the Post-Tertiary rocks of England.

The principal British elephant was the Mammoth, and his wanderings extend throughout the whole of Northern Europe and Siberia. Thousands of mammoth grinders, or teeth, are found in England, and thousands of mammoth tusks lie in Siberian soil. This great creature was commonly about twice as heavy as a modern elephant, and one-third as tall again.

A whole frozen mammoth was found in Siberia— a rare geological specimen! Even the flesh had been preserved; and when the enormous body was released from its ice-sepulchre, the bears and wolves ate it up. This particular mammoth was nine feet high and sixteen feet long, not counting the great curved tusks. It was covered with thick black bristles, about fourteen inches long, with thick hair four inches long, and with wool about one inch long. Such a coat might in general well fit the elephant for an Ice-Age, though even it seems not to have

ANIMALS OF THE TERTIARY AGE.

sufficed for the defence of the mammoth in question from Siberian frost.

How long he had lain in his ice-tomb no man can say. The 'very fresh' condition of the flesh and of the whole skeleton, would incline one to think that the time could scarcely be so long as some suppose.

Also, how he met his death we know not. The frost must indeed have seized with a sudden and mastering grasp upon his huge frame, thus perfectly to preserve it for later inspection.

Besides the Mammoth, the Woolly Rhinoceros flourished in Siberia—another instance of a usually tropic animal fitted to bear Arctic cold. The carcase of one has been found preserved in ice.

In North America the quadrupeds were as a rule less large than in Europe; yet the other continent had its fair share of big brutes.

Mastodon remains are found in America as well as in Britain and other parts of Europe. This enormous animal was a kind of elephant, peculiarly colossal and powerful in build. A mastodon skeleton may be seen in the Natural History Museum.* Another, found in America, is eleven feet high, with tusks twelve feet long.

* South Kensington.

There was also a huge brute,* somewhat approaching the modern sloth in kind—a tremendous massive slow creature, bigger than a rhinoceros. One such skeleton is eighteen feet long.

* Megatherium.

CHAPTER XXI.

THE AGE OF MAN.

'I have made the Earth, and created Man upon it.'—ISA. xlv. 12.

AMONG the uppermost animal-remains in the uppermost strata of the great Earth-crust Building, we meet at last with the remains of Man. Head of the Animal Kingdom, noblest of God's works, himself a Mammal, yet utterly superior to and separate from the whole brute creation, MAN crowns the ascending steps of the scale of Life upon Earth. We have now at length reached a time which is but as yesterday compared with the Ages before.

Different kinds of human relics are found, such as sometimes a human skeleton or a human bone; sometimes a bit of human workmanship; sometimes long-disused human dwellings; sometimes human tools of a more or less rough and early type.

The latter sign of man's life upon earth in any particular place should, however, be accepted with exceeding caution. The roughest and earliest flint tools of human workmanship approach so closely to the roughly chipped flints of Nature's workmanship, which exist by tens of thousands, that the one may often be mistaken for the other; and many so-called 'tools' have proved, or may yet prove, to be nothing of the kind.

In the course of long searching into such human remains as are to be found in the uppermost rocks, certain curious facts have been observed.

It is found that in each nation, as a rule, the earliest inhabitants, who were generally savages, or who were at least untaught and uncivilized, have made use of the simplest material lying ready to their hands, this material being *stone*. Later on they have begun to find out the uses of metal, and have learnt to fashion some sort of rough *bronze* weapons and implements. Later on still they have discovered the uses of *iron*.

The order of events has not, of course, been the same in every nation in the world. Still this particular order has been so often noticed, that the life of Man upon earth has been by common consent divided into three ages—The Stone-Age, the Bronze-Age, and the Iron-Age.

It must not be supposed that at any one time a Stone-Age was over all the earth, followed by an universal Bronze-Age, and then by an universal Iron-Age. Some countries which were very early peopled have left their Stone and Bronze Ages behind them long long ago, while others more lately peopled have not yet reached the Iron-Age.

We in England are living in a decidedly Iron-Age, having left far behind us our Stone and Bronze Ages. But a great many of the natives of the South Sea Islands, and of Central Africa, are still living in their Stone or Bronze Ages, and have no notion yet of the Iron. Again, the Stone and Bronze Ages in different parts of North America lasted up to some two hundred years ago, and then by a quick transition she passed rapidly on into an Iron-Age.

So the inhabitants of many countries travel through these stages, but not the inhabitants of all countries side by side.

Each country has its Historical Period, and its Pre-Historical Period, and its Geological Period. About the Historical Period we have more or less clear accounts handed down in writing from generation to generation. Sometimes we have the full history of a country from the time when it was first peopled; but more often the Pre-Historical Period means a time when men lived there, but about which little or no history has come down to us. Sometimes geology can take up the thread where history fails, and can give us dimly a few scattered facts as to the life of man in a country before historic times.

As we go back in English History, we have tolerably clear light until we reach Saxon days. Then, as we pass on, a kind of morning twilight remains through a few hundred years, till we reach the days of Julius Cæsar. A little way further still we may grope our way in a dusky atmosphere, but soon all surrounding scenery is lost in black night, and written records fail. A few fitful legends, like the dancing *ignis fatuus*, rather help to lead us astray than to light us onward, and presently even these fail. Then it is, if not sooner, that we reach the Pre-Historical Period; and then it is that geology steps in and gives us a few hints as to a possible Stone-Age before.

Scottish history is lost in fog and darkness rather earlier, while in French history the light lasts a little longer. In Grecian history you may wander onward for a considerable distance further before night shrouds the landscape.

So there is a great difference between the nations. The historic period of the United States is but a thing of yesterday, compared with the historic period of Great Britain. The historic period of Great Britain is but a thing of yesterday, compared with the historic period of Egypt or of Assyria. In each of these different instances, where history fails, geology may take up the thread and tell us a little, a very little, about the age preceding. But all dates are in confusion and uncertainty where we have only the geological volume to depend upon.

One singular fact as to human remains is that they are actually found, in England and France, side by side with the bones of the mastodon, the cave-bear, the ancient hyæna, and other animals now no longer seen.

It is not safe to build too much upon this fact. As before stated, the upper layers of earth and rock may have been in many places so changed and disturbed by the action of great floods in later times, as very much to affect our power of rightly reading them. Still, the impression given by these bones being found together naturally is that the animals and men to which they belonged probably lived at about the same time.

There are caves in the south of England and in the south of France where very large numbers of bones have been discovered, not lying on the ground, but buried a little way below ground. Bones of hyænas, wolves, and bears are found in profusion; and in very rare instances, quite near the surface, a human bone, or even a human skeleton, is discovered among them.

It is not an absolute certainty, because a human bone is seen there, that the human being lived alongside with the other animals whose bones are in the same cave and at the same depth in the soil. Some who have carefully examined these caves believe it possible, from the singular manner in which the bones are mixed up and rolled together, that a strong flood may have poured through the cave, washing earth, and perhaps bones also, in from outside, and destroying all regularity of arrangement. Whether or no this was the case, the bare idea shows how little we can build upon appearances without further knowledge.

But suppose that when such further knowledge is obtained, we find stronger and stronger proof of men having really lived alongside with certain extinct animals?

Why not? We do not know at what date any one species of any one kind of animal did really die out.

We only know that they do not exist now, and that above certain rock-layers we have not yet alighted upon their remains. But any one kind of animal may have existed still, though in lessening numbers, long after the particular age of the last fossil or bone of that particular kind found by man.

Many specimens of these creatures *may* have lingered on past the time of Adam or into the times of the Patriarchs. Some have thought Job's description of Behemoth wonderfully suited to the huge mastodon—far more so than to the modern elephant; and Job is believed to have lived in a very early age of the history of man.

Moreover it is well known that animals have died out of existence in a much shorter period than since the days of the Patriarchs. The Dodo, a large bird, which in the seventeenth century was very plentiful in the islands of the Indian Ocean, has since become entirely extinct, not one solitary specimen remaining. This is but one instance among several.

Sometimes the Stone-Age is divided into three parts. The middle one of these, the Reindeer Era, is supposed to have been a second Ice-Age, a return of cold, severe, though less intense than the cold of the great Ice-Age, taking place after a period of comparative mildness.

All these later sub-divisions are, however, very uncertain. Different books give them differently, and the same plans or rules cannot be applied to all countries alike. The whole may have to be rearranged before many years are over, as advance is made in knowledge. Exceeding caution in decision, and patient willingness to wait for further information, are of the utmost importance in this science, even more than in other sciences, to keep us from falling into grave blunders, and to guard our feet from dangerous quagmires.

CHAPTER XXII.

THE TWO RECORDS.

'Lo, He that formeth the mountains, and createth the wind, and declareth unto man what is His thought, that maketh the morning darkness, and treadeth upon the high places of the earth, The Lord, the God of Hosts, is His Name.'—AMOS iv. 13.

A FEW more words, and the Second Part of my little book will be ended.

This is not, indeed, a religious work, but no volume on the subject of Geology can be fairly and honestly written without frequent reference to the Divine Architect of the great Earth-crust Building.

Some slight mention has already been made of the first chapter in Genesis, and the Days or Ages of Creation there described.

To reconcile each separate detail, as given in the Book of Divine Revelation and as given in the Book of Nature, is a matter not yet possible with our present imperfect knowledge.

Many explanations have indeed been offered of the Bible record of Creation, in connection with late scientific discoveries; and many attempts have been made to dovetail the one account in with the other. Any one of these may approach more or less closely to the true explanation, though in each some flaw may be discernible.

But what then? What if all explanations hitherto offered are more or less mistaken? What if the true clue to the perplexity, the real point of harmony, lies at so lofty a height as to be beyond the utmost stretch of human intellect?

Then be it so. The grand Truth of either record remains still unshaken.

For we must ever remember that these two records, the story told in Genesis and the story told in the Earth-crust, proceed both from the same Divine Author. They do but give different views of the same grand realities. Though our reading of the one may seem to conflict with our reading of the other, they cannot be in themselves at variance.

Moreover, the one Record may and should be used as a help to the understanding of the other. If certain new discoveries prove beyond a doubt that the commonly-received interpretation of certain Bible-words is wrong, then that interpretation may be given up, but the truth of the written Word stands untouched. There should, however, be the utmost caution shown in accepting new theories and explanations. Many, if left alone, soon die a natural death.

Many questions must remain for a while longer in uncertainty, and we must be content to have it so, knowing our own liability to misread and misinterpret both the Written Volume and the Rock-Record.

With regard to the Bible-Record of Creation, a few brief suggestions may be a help to some minds.

We do not know how far the inspired writer was led to give a narrative of events in the precise order in which they actually occurred. A certain free grouping of leading events, for the sake of brevity, within the limits of exact truth, is sometimes adopted in very short narratives. The writer appears here, as he passes rapidly on, to seize in each Day upon one or two leading points of interest—the prominent facts of the period. Many lesser points of interest and minor facts doubtless existed alongside; but with these he is not concerned. He gives the names of the topmost mountain-peaks in the scenery, and attempts no description of lower heights.

We do not know how far his language is literal and how far it is figurative. Both styles are largely used in the Bible.

We do not know whether that which he describes was told him in words, or was revealed to him in vision; and if the latter, whether it was by one vision or by a succession of visions.

We do not know how far he was led to describe things simply as they would have appeared to an observer, standing on the unfinished earth during the Creation-Days, — a style not unfrequently employed in the Bible. This would tend to explain certain difficulties, such as the mention of light on the First Day, but of the sun and moon not sooner than the Fourth Day.

We do not know whether, as some think, the Days may mean vast periods or ages of time, or whether, as others think, they may rather refer to an *appearance* of days and nights, caused by a succession of visions sent to Moses, each in turn dawning, brightening, and vanishing in darkness.

It has been the opinion of many that between the first and second verses in Genesis a vast interval of time is, or may be, passed over.

Some have even held that the countless ages of Geologic history may have been contained in that interval, followed by universal destruction of life, by a period of chaos, and by Days of new creation.

Others again have been struck with certain broad outlines of remarkable agreement in the Days and the Ages of the two Records: first, as to the gradual preparation of the earth; secondly, as to the gradual development of life upon earth, proceeding upwards step by step from lower to higher forms.*

The above are some among the many suggestions offered by different writers. We shall be wise to hold our thoughts free, and to wait for fuller knowledge of the deep meanings underlying the brief Bible narrative. When both records are fully understood, we shall be amazed at the majestic completeness of the two combined, at the wondrous simplicity of that which now seems to us complex and mysterious. By-and-by we shall have power to grasp the whole. At

---

* One hint may be offered here with regard to the creation of 'great whales,' spoken of in Gen. i. 21, as preceding the creation of Mammals, which was in turn followed by the creation of Man. The Hebrew word there translated 'Whales,' and elsewhere translated 'Dragons' (see Psa. lxxiv. 13; xci. 13; cxlviii. 7, etc.), is derived from the verb *tanan, to extend*, and may signify any *long* sea or land animals, serpents, crocodiles, etc., as much as whales. Also the Greek word, κητη, used in the Septuagint translation of the same, does not necessarily mean whales only, but any huge sea-monsters. A recollection of the Age of Reptiles naturally occurs to mind.

present we know only that—'HE SPAKE, AND IT WAS DONE; HE COMMANDED, AND IT STOOD FAST.'

A more easy task lies before us in the third and last section of this little book. We have learnt, first, the simple alphabet of Geology. We have read, secondly, a brief sketch of the history written in earth's rocks. We have to gather, thirdly, from the events going on daily around us and in all parts of the world, certain facts which may help us to a better understanding of the great Rock History—to view, as it were, the Past in the light of the Present.

For though we may in a sense talk of the Earth-Crust Building as finished—the lower, middle, and upper stories all complete—yet it is so only in a sense. Still the same busy workers are ceaselessly employed, ever pulling down and building up, ever making alterations, ever taking away here and adding there.

Much more attention has thus far been given to the working of the great agent, Water, than to the working of the great fellow-agent, Fire. This could hardly be otherwise, since it is in the Water-made rocks, and not in the Fire-made rocks, that we find the written record of the life of plants and animals upon earth.

The fact should not, however, be lost sight of that,

side by side with the writing of the life-record in the water-built rocks, the formation of fire-made rocks was going on in different places. There are such rocks, sometimes arranged in layers and sometimes not, believed to belong to each different period in Geology. But it is much more difficult to fix the comparative ages of fire-made than of water-made rocks. By their 'ages,' I mean ages counted in periods, not in centuries.

In the Third Part, somewhat fuller attention will be given to the working of the mighty underground agent, Fire, though want of space will forbid full details on any subject.

A word of caution is perhaps advisable at the outset.

We may fairly and rightly examine the working of the different forces as seen in the present day, and may learn much from that working about their action in past ages.

But there is a danger lest this mode of reasoning be pushed too far.

There have been men of powerful intellect who, long and earnestly studying thus, have arrived at the belief, that precisely as these powers are now observed to work, so they always have worked through countless ages past. In other words, they hold that throughout the vast periods of Geological history, rocks have never been more rapidly wasted, river-deltas have never been more rapidly formed, coral, chalk and limestone have never been more rapidly made, earthquakes have never been mightier, land has never risen or sunk more extensively and rapidly than the speed with which these things have been known to take place within the last few hundred years.

That such has been the case we can, to say the least, possess no certainty. The fact that a particular thing has existed under particular conditions during so many hundreds or thousands of years, affords no proof whatever that it so existed during preceding ages. The one does not follow as a necessity upon the other. We may guess, may suppose, may imagine, but we cannot KNOW it to have been thus.

We see about us certain powers incessantly at work. Water wears away land. Fiery heat surges underground. Coral islands and reefs are slowly built. Limestone is gradually formed. These things go on now, and these things have gone on in the past. It is, however, no easy matter to say how fast they go on now, since the rate of waste, or of growth, or of change, is seldom the same in two places. And even if we could fix on any particular rate for each, as generally not far wrong in the present, that would be no sure guide as to the past. Differing circumstances in the way of climate, atmosphere, land, ocean-currents, underground forces, may each or all have affected in a marked degree the speed of alterations on the earth's surface through earlier times.

For after all, when we talk of judging the past from the present, what is that 'present?' The life of man upon earth is but a thing of yesterday, compared with the ages preceding. What do we know of the mighty changes, the vast upheavals, the great dislocations, the tremendous destructions of life, the wondrous times of renewal, which may have taken place —ay, and may equally have taken place either slowly or rapidly, gradually or suddenly? God may have had His very different modes of working in different periods, whether known or unknown to us.

Even supposing that the forces of Nature may have continued the same in intensity during hundreds of thousands of years past—a question which cannot be decided with certainty—is it conceivable that they should have been the same in yet earlier ages, before the earth had parted with its heat so far as it has since done?

For there was a time, as we believe, when our earth was a sun—small indeed compared with the great central orb of our system, yet a true burning sun, a tiny star shining by its own light. Looking back to that far-off time, remembering the countless ages between, and the mighty changes involved, this 'uniformity theory,' as it is called, seems a thing incredible.

Thus, here, as in other matters, it befits us to be cautious, to be humble, to be content to await fuller knowledge.

On the page following will be found a list of the Geological Periods and Ages, somewhat more full than the short Table of Strata already given.* It should perhaps be mentioned that the distinction between Periods and Ages, kept up generally for the sake of clearness throughout this volume, is not observed as a constant rule by writers on Geology. The two terms are often used interchangeably, not to say confusedly.

* Page 60.

THE AGES OF GEOLOGY.

I. PRIMARY (FIRST) PERIOD;

OR PALEOZOIC (ANCIENT-LIFE) PERIOD.

1. LAURENTIAN AGE; so named from rocks found near the River Lawrence.
   First and Simplest Forms of Animal Life.

2. CAMBRIAN AGE; so named from rocks found in Wales; sometimes counted as part of the Silurian.
   Age of Invertebrates, or Boneless Lower Animals.

3. SILURIAN AGE; so named from Silures, ancient name for a tribe in Wales, where these rocks also are found.
   Age of Invertebrates, or Boneless Lower Animals, continued.

4. DEVONIAN AGE; also known in part as Old Red Sandstone. Called Devonian because largely visible in Devonshire.
   Age of Fishes, or First Backboned Animals.

5. CARBONIFEROUS AGE; rocks found in England, France, America, and elsewhere.
Age of Coal, Age of Ancient Forests, and Age of Amphibians.

6. PERMIAN AGE; so named from Perm, a Russian province, where these rocks are found; sometimes classed as one with, and sometimes as following after, the Carboniferous Age.
Close of the *Forest Age.*

II. SECONDARY (SECOND) PERIOD;

OR MESOZOIC (MIDDLE-LIFE) PERIOD.

1. TRIASSIC AGE: so named by German writers, because in Germany these rocks are divided into a Triple group.
Age of Reptiles.

2. JURASSIC AGE; so named from the Jura Mountains, where these rocks are visible.
Age of Reptiles, continued.

3. CRETACEOUS AGE; so named from a Latin word *creta,* chalk.
Age of Chalk; also *Later Age of Rhizopods;* also *Age of Reptiles,* continued; also time of first appearance of *Mammals* and *Flowering Plants.*

III. TERTIARY (THIRD) PERIOD;
OR CAINOZOIC (NEW-LIFE) PERIOD.

1. EOCENE AGE; from Greek words, $ηως$, dawn, and $καινος$, new. *Shells*, very few of living species. *Age of Mammals.*

2. MIOCENE AGE; from $μειον$, less, and $καινος$, new. *Shells*, less than half the number found, of living species.
Age of Mammals, continued.

3. PLIOCENE AGE; from $πλειον$, more, and $καινος$, new. *Shells*, more than half the number found, of living species.
Age of Mammals, continued.

4. PLEISTOCENE AGE; from $πλεῖστος$, most, and $καινος$, new. *Shells*, the greater number found, of living species.
Age of Mammals, continued.

POST-TERTIARY AGE; sometimes included in the Tertiary, sometimes counted as following after.

1. *Post-Pliocene:*
Age of Mammals, continued.

2. *Recent:*
Age of Man.

In America this Post-Tertiary (or After-Third) is called the Quaternary Age; and is divided into—

I. The Glacial Age, or Age of the Drift;

II. The Champlain Age; divided again into Diluvian and Alluvian Epochs;

III. The Recent or Terrace Age; divided again into Reindeer and Modern Era.

But these particular arrangements and subdivisions are arbitrary, and must remain subject to much future alteration.

PART III.

THE PAST IN THE LIGHT OF THE PRESENT.

CHAPTER XXIII.

RIVERS.

'In Thine hand is power and might.'—1 CHRON. xxix. 12.

MUCH has been already said about the work performed in the world by running water. This work may be divided into three distinct parts. First, water wastes, breaks up, crumbles, or wears away land. Secondly, it carries the wasted material, whether rocks, stones, pebbles, sand, or soil, towards the ocean. Thirdly, it drops or deposits that material on the sea-bottom.

Most of the valleys, ravines, gorges, and clefts in the world have been more or less formed by the action of running water—sometimes partly, sometimes entirely.

When you see a stream rushing swiftly down a mountain-side, you are inclined to think of that stream as a fixed part of the scenery—at one time, indeed, more full than at another time, but remaining through centuries the same.

Yet this is far from being the case. The stream is busily engaged in cutting out for itself a pathway deeper and deeper into the earth or through the rock, whichever may form its bed. Each river, each stream, each brook in the world is doing this work. And very wonderful it is to see how the hardest rock is worn away by the soft water which runs perpetually over it.

Water is made of the two gases, Oxygen and Hydrogen. Perfectly pure water would have little or no power in wearing away rock, but it almost always contains a certain amount of the powerful Carbonic Acid Gas. This gas, so needful to the structure of plants and animals, yet so fatal to animal-life, has a singular power of eating away rock, and causing it to crumble beneath the flowing water.

Water when warm dissolves hard substances much more rapidly than when cold. Warm water in a natural state is now only to be seen in certain places, but in the earlier ages of the world's history it *may* have been much more common. This is one of the 'may-bes,' which render utterly uncertain all attempted calculations of time in those early ages.

The movements of stones and sand in running water help that water in its work. Every rock that grinds against another rock, every stone that is washed against the bank, every grain of sand that rubs in passing against a boulder, does its little share in the task of ' wearing away.'

At the Linn of Quoich, which is part of the River Dee, a singular instance may be seen of the cutting and carving power of running water. A neat round hollow in the hard rock, close to the stream—laid bare in dry seasons, but overflowed when the water is high—has long existed, called in the neighbourhood 'The Earl of Mar's Punch-bowl.'

In past days this hollow was like a huge round cup in the rock, with a solid rock bottom. But the waters which first made the hole did not stop there. Still they went on washing round and round, carrying pebbles round and round with them, and still the restless waters and the hard stones continued their 'wearing away' work, till at last the bottom was quite gone, and now the former 'bowl' is a deep round hole, with water filling it from below. It was probably begun, in the first instance, by the simple rocking to and fro of a boulder on the bed of the stream. A slight hollow being thus formed, the water would soon take to a circular motion in the hollow, and the movement of stones with the water would gradually accomplish the rest.

Such holes are sometimes called Pot-holes. There is a large one in America, named 'The Basin,' fifteen feet deep and over twenty feet across. Some very large holes of this description may also be seen at Lucerne, displayed as 'Glacier-holes,' but more probably in the first instance fashioned by the action of running water and circling stones.

A stream coursing quietly over an almost level plain has no great wasting power. It is when water descends a steep height that the work goes on most quickly. A waterfall, or a torrent on a mountain-side, cuts its way rapidly backwards.

Sometimes a powerful mountain torrent may be seen rushing down a narrow gorge, with high rocky precipices rising steeply on either side. In such a case you may be pretty sure that, whether or no the work of gorge-forming *began* thus, the water has had a great deal to do with carrying it on. Once upon a time the stream probably ran at a much higher level, a good deal nearer to the tops of the cliffs than now, and through centuries past it has been gradually eating its way down to its present level. The passage of the foaming Reuss through the rocky pass of St. Gothard, in the neighbourhood of the Pont du Diable, is a good example of this.

It is probable, however, that most of the larger valleys have not been formed by the action of water alone, but in a greater or less degree by the effects of underground disturbances, through earthquakes or earth-splittings.

Some instances of very rapid valley-forming through water-action have been seen.

On the Vispbach, in 1857, a sudden landslip laid bare an underground spring of water, unknown before to have been in existence. Henceforth a stream flowed down the mountain-side from the unearthed spring, and this stream immediately began the work of cutting for itself a channel. In the course of three years it dug a 'gully,' as described by an eye-witness. Eight years later, when the same eye-witness went to the spot, he found that the stream had deepened and widened the little gap to such a degree, that a vineyard, which before the opening out of the spring had been unbroken, was cut in sunder by a chasm over forty yards in width, and at its shallowest part some fifteen feet in depth.

It has been found in America that the levelling of ancient forests often results in new valleys being formed. Near Georgia a certain forest was thus cut down, uncovering the clay soil, which the heat of the sun thereupon cracked in many places. Some of the cracks were three feet deep. When the rains set in, a rush of water taking its course through the largest crack rapidly deepened it, and at one end steadily wore a way backwards. The crack grew into a chasm, widening, lengthening, and invading the high-road which lay near at hand. In the course of twenty years this new little valley increased, till it measured three hundred yards in length, fifty-five feet in depth, and at its broadest one hundred and eighty feet in width. Another such gorge, twice as large, was made in Brazil in forty years.

The material worn away in these cases was soft. Where solid rock is concerned the work is necessarily slower; yet even here the speed is often greater than one would expect.

At the base of the volcano, Etna, there are vast quantities of lava, poured at one time and another out of the mountain. The lava from one great eruption flowed down into the valley of the River Simeto, filling up its channel for some distance, and passing in great masses to the other side. This particular outbreak is believed to have taken place in the year 1603. In the course of about two centuries the river cut for itself, through this lava—a peculiarly hard and firm kind,—a passage, from forty to fifty feet in depth, and in parts several hundred feet wide.

The mighty Niagara Fall must have been for ages past slowly eating its way backwards in the rocks over which it flows; so that the spot where the fall now takes place is not the same spot where it used to take place. This is more or less what happens with all water-falls, and certainly not least so with the great falls of the Niagara.

A good many attempts have been made to fix the rate at which the Niagara works its way backward. One supposes it to be at the speed of a foot each year, while another suggests that it may be only three feet in each century. But careful observation alone, through long periods, could supply us with any means of truly answering this question; and careful observation in such matters is but a thing of yesterday. Even if the present rate of wear were clearly known, this would form no safe guide as to the past. At every step in the path of the retreating flood the character of the materials to be worn away must have varied, their hardness or softness greatly affecting the speed of their destruction.

That this work does actually go on may, however, be plainly seen. Every year there are changes in the shape of the channel; and huge rock-fragments are being perpetually broken off and dashed into the foaming waters below.

When speaking of the work of rivers in carving out valleys, it should be borne in mind that a river acts in two ways. First, there is its regular daily work in its narrow channel, lasting usually all the year round. Secondly, there are the flood-times, when the river spreads over a much wider bed, and often does much damage. In England such floods, though injurious, are comparatively quiet; but in some countries river-floods are sudden, widespread, and fearfully destructive.

A river may therefore be said to have two beds— its narrow constant bed, and its wide occasional bed. The latter is called often an *alluvial plain*. When we talk of a river overflowing its banks, we really mean that it is running in its wide bed instead of its narrow one.

The work of wasting and wearing away, which is done by rivers and torrents and cascades on a large scale, is done by every little brook and streamlet on a small scale. If rivers carve out valleys, brooks hollow out dells. The materials which are crumbled away in the making of these valleys and dells, are borne seaward—first carried by the brooks down to the rivers, and then carried by the rivers down to the ocean. The amount of land thus torn yearly from the continents is past calculation.

CHAPTER XXIV.

WATERS.

'This great and wide sea.'—PSA. civ. 25.

NOT rivers and streams alone do the work of wasting away land. Heavy rains, underground water-flows, ocean-waves and tides and currents—all take an active share in the same.

Even in England long-continued rains may cause much material to be carried off—witness the brown muddy streams seen to flow at such a time, the brown tint coming simply from the earth which the water is stealing from the land.

But it is in foreign countries that the full effects of rain may be observed.

At Chirapoonjee in Bengal, high up among the mountains, the rainfall is tremendous. Thirty inches have been known to fall in twenty-four hours, and through the whole year the amount is more than twenty times as much as that which falls in one year in England. The deluge of water down the mountain-sides, during the rains, destroys vegetation, bears away soil, and makes a wild waste of what might otherwise be a richly-wooded land.

In another part, near the Sikkim Mountains, the downpour of water in the rainy season is such that rivers have been known to rise twelve feet in twelve hours. The rush of torrents, the sound of falling trees, the crash of boulders dashed one against another by stormy cataracts, are described as sometimes continuing night and day unceasingly.

The sudden and violent rains of the tropics thus tear away earth, and grind rocks and stones into powder, far more rapidly than the rains of temperate lands can do.

There are streams and rivers underground as well as above ground. When such a stream breaks out of the side of a hill, we call it a 'spring.' Underground channels and caverns are hollowed out by the flowing waters, with the help of the Carbonic Acid Gas which they contain. All rain-water which does not run to the ocean through streams and rivers sinks into the ground, joins the waters there, and in time generally finds its way to the ocean.

The direction taken by underground streams is much affected by the kind of soil. Loose soft gravelly or sandy soils allow water free passage, but tough clay or hard rock act as barriers.

Sometimes a large reservoir of water will collect over one stratum of rock or clay and under another, unable to find an outlet. If the water has found its way there from a greater height, the pressure of other water trying to flow in from behind will make it ready to take advantage of any opening that may occur.

In such cases as this wells are often made. Men bore down from above, lowering a slender tube as they bore, through several kinds of soils, till the said tube passes through the upper clay and reaches the imprisoned water. Instantly the water rushes up the tube, glad, as it were, to find an outlet.

Wells formed thus are called 'Artesian,' because the plan was first tried in Artois. They are often found useful in supplying water to a neighbourhood, where the amount within reach would otherwise be scanty.

It is impossible to tell, with exactness, where these reservoirs are without boring, and many attempts have therefore been made in vain. In other instances, however, the toil has been amply repaid.

For the success of such a well, it is needful not only that there should be the reservoir of imprisoned water, but also that the water should have flowed down from a higher level, the way by which it has come being still open. If you were to make your boring at the top of a hill, with no higher ground near, then, even though the tube should reach an underground reservoir, no water would rise, since there would be no pressure of water behind to force it to do so.

In a well bored at Tours, the rush of water was so great that it rose, like a fountain, to the height of thirty-two feet above ground. In another at Grenoble, none appeared till the tube had reached a depth of two thousand feet. Then it came in good earnest— soft warm water, pouring steadily up from lower regions, at the rate of half a million gallons in each twenty-four hours. Such wells have sometimes been successful even in deserts.

Occasionally the flow of water is found after a while to lessen, proving that the reservoir is not a very large one.

All around the shores of the British Isles, not to speak of other lands, the ocean-waves are beating, beating perpetually, wearing away her cliffs, eating into her shores.

Just as, with the rivers, the chief waste is on the hill-sides, and not on level plains, so also with the sea. The waves have comparatively little power to wash away the soft flat sands. It is upon the bold cliffs— not only cliffs of soft chalk, but of hard rock also— that their power is chiefly shown.

There is much less wear and tear on the Mediterranean coasts, where the tides are so slight as to be almost nothing, than on British shores, where the rise and fall of the tides are great. Yet, even in the Mediterranean the waters are doing their work, and every storm leaves its traces.

In the Shetland Islands, exposed as they are to the full force of the broad Atlantic, the action of the waves is forcibly displayed in the fantastic rock-groupings, the caves and arches, the columns and pinnacles, the needles and obelisks, formed by the wear of the hard rocks under the incessant beating of the surges. Much the same may also be seen along parts of the exposed west coast of Scotland.

At the Bell-Rock Lighthouse, the wonderful strength of ocean-waves is seen perhaps as clearly as anywhere. Stones over two tons in weight have often been flung upon the rock by the billows in their wild gambols. While the lighthouse was being built, six large granite blocks were placed upon the reef ready for use, and all the six were heaved by the waves over a ledge to a distance of more than twelve paces.

On the east coast of England, the wear of land is markedly shown. There were, once upon a time, certain Yorkshire towns or villages, named Auburn, Hartburn, and Hyde, but they are to be seen no longer. Sandbanks, overflowed by the sea, lie over their former sites.

Along Norfolk shores, the chalk-cliffs are crumbling steadily away before the waves. In the year 1805, an inn was built in Sherringham, at a certain distance from the cliff-edge. It was known well that the sea, eating away the land, must in time approach the inn. Observations had been carefully made, and the wearing away was believed to be at the rate of less than one yard each year. It was calculated, therefore, that the inn might be considered safe for seventy years to come; yet a very short time proved this calculation to be wrong. In 1805, full fifty yards lay clear between the house and the sea. In 1829, only fourteen years later, seventeen yards of land were already swept away, and only a small garden divided the inn from the devouring ocean. An instance this of how easily mistakes may be made in any such attempted calculations.

Also in Norfolk whole towns have been demolished.

Ancient Cromer lies beneath the Ocean. Shipden, Wimpwell, Eccles,* have been slowly swallowed. One town and another, on the east coast, is compelled to beat a gradual retreat before the enemy, building more and more inland, yielding one house after another to the encroaching waters.

In other parts the same is seen,—markedly so with the crumbling white chalk cliffs of the south coast of England.

When Queen Elizabeth reigned, Brighton was built upon the same belt of land, where now the chain-pier runs out into deep water. Measures have been taken to check these inroads of the sea; but for centuries Brighton went backwards, step by step, before the ocean.

Between Hastings and Eastbourne the shore-line has long receded steadily before the waves, and a haven which once existed in Pevensey Bay is now filled up with shingle.

* Since writing this chapter, I have visited Norfolk, and have seen the church-tower of Eccles still standing upon the sandy beach, the body of the building having been almost entirely washed away. A year ago the aisles could be traced; now they are gone. The village of Eccles lies beneath the present shore. An eye-witness told me that, many years ago, a heavy storm once carried away so much of the sand, as to leave exposed the village site,—the foundations of the houses, and even the cart-ruts in the streets, being plainly visible.

In some parts the yearly waste has been as much as seven feet of land. At the neighbouring promontory, Beachey Head, a great fall of material took place in 1813,—a mass of chalk, three hundred feet long and seventy broad, descending with a mighty crash to the shore below.

Somewhat to the west of Newhaven, there are remains of an ancient earthwork, supposed to have been British. The greater part of this entrenchment has been carried away by the waves. Two other ancient camps, one near Seaford, and one near Eastbourne, have been in like manner partly destroyed. The same is seen in other parts of England.

It is said that during some eighty years no less than twenty distinct inundations of parts of the coast of Sussex took place, permanently overwhelming tracts of land, which varied in extent from twenty to four hundred acres.

Examples more or less like in kind might be brought forward, as to the coasts of Holland, as to other European countries, as to America. But enough has been said to show the power of the ocean, now and through ages past, in wearing away firm land.

CHAPTER XXV.

DELTAS.

'He hath founded it upon the seas, and established it upon the floods.'—PSA. xxiv. 2.

THE wearing away of rock and earth is not the only work done by water in the world. For while it pulls down, it also builds up; while it wastes, it also heaps together. The material which it steals from the land in one place, it often adds to the land in another place.

One has well said of a mountain-torrent: 'It lays down what it will remove, and removes what it has laid down.' This, which is true of every torrent, is true of all streams and rivers, nay, of the very ocean itself.

The work of building up land is most plainly to be seen at the mouths of rivers. Deposits on the ocean-bottom, at a great distance from shore, doubtless take place constantly, but we cannot see this for ourselves; whereas the growth of land at a river's mouth may easily be observed.

When a river, laden with sand and earth which it has stolen from the land, reaches the sea, the speed of the flowing water is suddenly checked by the incoming waves. The weight which the river was able to carry, while moving quickly, it can no longer support, and sand and earth sink to the bottom, forming there in layers.

This is the way in which sand-bars or mud-banks are made. In the case of a small river, the deposit takes place very near the shore, sometimes almost choking up the mouth of the stream. The larger and more powerful the river, the farther out to sea will the materials be carried before they are dropped.

This is the way also in which river-deltas are made. The name Delta is given to the tract about the mouth of a large river, because of its likeness in shape to the Greek letter of that name Δ. The river, after flowing long as a single stream, divides into two or more streams, branching off and reaching the sea by different channels.

The low lands lying between these different channels have been slowly built up by the river out of the materials which it has stolen on its course. The whole of these lands, from the spot where first the river separates into two down to the ocean, is called the Delta.

Many changes take place in the deltas of great rivers. Now one arm and now another becomes quite choked up with sand or mud, and the water ceases to flow there, taking to another channel.

There are three kinds of deltas. First, those which are formed by rivers flowing into lakes. Secondly, those which are formed by rivers flowing into inland seas, where there is almost no tide. Thirdly, those which are formed by rivers flowing into the ocean where there are full tides. The first two are much alike, except that the kinds of animal-remains found in them differ; being in the one case those of fresh-water creatures, and in the other case those of salt-water creatures.

The Rhone, passing through the Lake of Geneva, gives us an example of a Lake-Delta.

Building up of new land in the lake has continued there through ages. Laden with sediment torn from the crumbling mountains, the river enters the lake at one end, drops its material layer upon layer, passes through the whole length of the lake, and flows out at the other end, pure and clear, having left a load of mud behind.

The land thus formed grows steadily, reaching farther and farther into the lake. A certain town, Port Vallais by name, which in the days of the Romans was close to the water's edge, is now one mile and a half inland, showing that about one mile and a half of land has been built by the river in the course of the last eight hundred years. Before that date five or six miles of the delta had been built.

The Rhone does not long remain pure after its passage through the Lake of Geneva. It is speedily joined by the Arve, heavily laden with sand and other materials from the glaciers of Mont Blanc. More rivers join her later, bearing sandy and muddy donations from the Alps of Dauphiny. When at length the Rhone enters the Mediterranean, its powerful stream stains the blue waters for a distance of six or seven miles, as it drops once more its later burden, this time in the sea.

The Rivers Po and Adige, flowing into the Adriatic, give an example of rapid land-building. From fear of their inundations at flood-seasons, the inhabitants of the country have long built embankments to close them in. The rivers, being thus unable to spread themselves over their wider beds, and to make deposits on the low lands around, are compelled to carry nearly all their sediment to the sea.

The growth of land there is consequently much more rapid than if there were no such embankments. It would have been more so still, but for the fact that the shores are, and have long been, slowly sinking.

Even thus, however, the increase of new river-built land along the coast, for a distance of one hundred miles, has been within the last two thousand years as much as from two to twenty miles in breadth. The town Adria, which in the time of Augustus borderedt he sea, now stands some twenty Italian miles inland.

The delta of the Nile is one of slow growth, but it may have been much faster in early ages, before it reached so far out into the sea as to be swept by the strong current which coasts the north of Africa. Much newly-formed land is from time to time carried away by this stream; and here again, as in North Italy, the gradual sinking of the shores prevents the more rapid apparent growth of the delta.

Moreover, although the Nile brings much mud and sand from the interior of Africa, a large proportion of the material is dropped upon Egypt in the flood-seasons. But for this Egypt would be a barren land indeed. Each flood-season, when the river overflows the flat lands around, it places a new thin film of earth upon the layers of centuries before, thus ever deepening the soil. Then, bearing on only the lightest and finest particles through the rest of its course, it reaches the ocean, and there drops them in some part of the many-armed delta, or carries them yet further, to discolour the blue Mediterranean through thirty or forty miles.

The Nile stands alone among rivers in the singular fact that, during the last fifteen hundred miles of its journey, it is joined by no tributary stream. The delta of the Nile is at its base, or where it joins the sea, about two hundred miles wide.

Two great Indian rivers, the Ganges and the Brahmapootra, offer another good instance of delta-building.

Coming from almost opposite directions, they meet and mingle, so that one vast delta serves them both—in size more than double that of the Nile. This 'great delta of Bengal,' as it is called, is in part made up of a bewildering maze of large and small streams, some filled with salt water from the inflowing sea, and some with fresh water from the outflowing rivers. The portion known as the Sunderbunds, a tiger-infested wilderness, is alone as large as Wales.

Perpetual changes take place in this immense delta. One season is a time of floods, and masses of new land are swept out to sea. Again, the ocean rushing in, carves out fresh channels in the low muddy banks. Or new islands are rapidly formed, only to be as rapidly destroyed. In one spot, no less than twenty-five thousand square miles of delta-land were carried away in the course of a few years.

Early in the present century, a new island was formed, about four miles out to sea, near the mouth of the Hooghly—one of the channels in this delta. It grew to a length of two miles and a half, and houses were built upon it. In 1823 there came a tremendous gale, and the island was cut into two smaller islands. A few years more, and the former island had been worn away to a mere low sand-bank; about half a mile long.

These two great rivers are, in fact, ever building up new land in and about the delta, while the ocean is ever seeking to destroy that which the rivers have built.

One more instance of a delta may be given in that of the mighty Mississippi. This great stream, if measured with its windings, has a length of three thousand miles, and the land which it drains is over half the size of all Europe.

The Delta of the Mississippi is about two hundred and forty miles across in one direction, and about one hundred and forty miles across in the other. It covers more than twelve thousand square miles.

The river travels chiefly in great bends, making sand-banks at each bend; often changing its course; frequently building up new land, and as frequently washing away what it has built. Enormous quantities of trees are carried down the stream, and dropped in the delta; and no doubt many animal-remains also are imbedded there in the low banks. The great force of the river bears it on as a fresh-water stream to a distance of more than twelve miles into the ocean.

Some attempts have been made to calculate or to guess at the time probably occupied in the forming of this vast delta.

A certain quantity of sediment was fixed upon as that which the river was supposed to bring down each year to the delta. It was also supposed that this quantity had always been about the same, every year, through centuries and ages past. It was then calculated that—*if* this quantity were correct, and *if* the amount of sediment had always been equal—the delta must have taken about sixty-seven thousand years to form.

A little later, however, it was found that the quantity of material brought down yearly by the river was very much more than had been first imagined. So the calculation had to be made over again ; and this time it was decided that the delta-lands might have been built in the course of thirty-three thousand five hundred years.

Whether these figures will have to be halved again remains to be seen. That the time was long, very long, appears highly probable. But how long it lasted man cannot say.

Indeed, apart from other difficulties in finding out the age of a delta-formation, the uncertainty as to how far land thereabouts may have risen or sunk in past ages renders almost useless any such attempted calculation.

Before quitting the subject of water-action, a little illustration from present days may be given, in explanation of the fossil rain-drop marks and fossil foot-prints so often mentioned in earlier chapters.

On the borders of the Bay of Fundy, in Nova Scotia, there are broad mud-flats lying within reach of the tides.

The rise of the tide there amounts to more than one hundred feet. Parts of these mud-flats are covered and uncovered every day ; but other parts, farther off from the sea, are only reached by the highest spring-tides, and for a week, or even a fortnight at a time, they remain dry and untouched by the waves.

It may be easily imagined how, in these upper reaches of mud, the marks of rain-drops or the footprints of passing animals, if received immediately after a high spring-tide while the mud is still soft, would have time to harden into a permanent shape before the next high tide came, a fortnight later, to deposit another layer of sand.

A traveller, passing the spot in question, afterwards described the mud near the sea as being too soft to retain impressions, and the mud far away from the sea as being too hard to receive them. But between these two he discovered a belt of soft mud, just of the right consistency to take and to keep markings. On splitting open slabs of this mud, he found prints or casts of rain-drops hardened into lower layers of the mud, afterwards covered over by later sand-droppings out of the waters.

CHAPTER XXVI.

GLACIERS.

'He casteth forth his ice like morsels; who can stand before His cold?'—PSA. cxlvii. 17.

THE work which Glaciers are believed to have done in past times has been already more than once described.

In the present chapter we have not so much to think about any one particular Ice-Age, as to gather from glaciers and icebergs of the present day certain facts which may help us to a clearer understanding of the past.

A glacier, as before explained, is simply a river of ice; not fed, like a river of water, by rains and springs, but by masses of snow and freezing mist in high mountain-regions. Also, just as a common river is fed by lesser streams, so a large glacier is fed by lesser glaciers.

The flow of a glacier is in many respects like that of a river. It usually follows the course of a valley. It travels faster in summer than in winter, and nearly as fast by night as by day. In the ice of a glacier, as in the water of a river, the movement is quicker at the surface than down below, and quicker at the middle than at the sides. The reason for this is that the bottom and sides, both of a water-river and of an ice-river, are retarded or kept back by the rubbing of the bed and banks.

Moreover, a glacier is like an ordinary river in the manner in which it suits itself to the shape of its bed, narrowing or widening according to need. A glacier is known to spread itself over a bed two thousand yards wide, and then to press through a gorge only nine hundred yards wide, still continuing its steady onward march. A glacier will pass round a bend, like a river; and like a river also it has its occasional cataracts or ice-falls, where masses of ice, in slow succession, plunge over a precipice. A glacier is, however, unlike a river, in the fact that it not only moves downhill and along level ground, but sometimes for a space will even travel *up* a gentle slope.

All these particulars make the subject of glaciers a mysterious one. A great many theories are put forward to explain them.

Some think that the onward motion of glacier-ice is caused by the earth's attraction; just as water is thus caused to run downhill. Some think that it moves only because the ice behind keeps pressing it on. Others think that the heat of the sun causes it to travel along the ground.

Ice is in a measure elastic, not hard and stiff like rock; and it is believed that the rapidity with which the masses of ice break and freeze together again has somewhat to do with the explanation of the mystery.

Icebergs are large masses or mountains of ice, which have snapped off from the bottom or foot of some enormous glacier, and have floated away on the sea.

Icebergs are often of a very great size. Whatever quantity of ice is seen to rise above the water in an iceberg, there is always about eight times as much below the water. You will understand this better if you float a small lump of ice in a basin, and notice what a small proportion of the whole lump remains out of the water.

The weight and force of these floating ice-mountains is sometimes terrific. Many a ship, caught between two of them, or between an iceberg and a grounded ice-field, has been crushed, as an egg-shell

FLOATING ICEBERG.

may be crushed between two fingers of a man's hand.

In England and Scotland glaciers no longer exist. They are found on the high mountains of the Continent, in the northern parts of Europe and North America, and also in Antarctic lands.

One of the principal Swiss glacier-districts is that of Mont Blanc. The vast snow-fields of the summit give rise to glacier after glacier, creeping slowly down each chief valley, the long tongues of ice reaching far below the usual 'snow-line;' and when at length they melt, sending on rushing streams laden with earth and stones.

The stones, many of them scratched and scored from being dragged along over the rocky bed of the glacier, are quickly dropped in heaps, forming the 'terminal moraine;' but the muddy turbid stream flows down the mountain-side, until it joins some ocean-bound river. The north-western glaciers of the Mont Blanc district send their streams into the River Arve, spoken of in the last chapter, while the south-eastern feed the Doire.

The great Mer de Glace, or Sea of Ice, so often described by travellers to Mont Blanc, is made up of several of these glaciers in the upper part of their course. The ice of the Col du Géant, travelling slowly down the Mer de Glace, does not reach the further end in less than one hundred and twenty years.

Some of the Swiss glaciers are as much as six hundred feet deep, twenty or thirty miles long, and two or three miles wide; but this is not common.

After all, the grandest of Swiss glaciers sink into nothing, when we leave them behind and wander in thought to the dreary wastes of Greenland. There it is that we may learn most about the possibility of a past Ice-Age upon the earth.

Little of Greenland is known beyond the strip of habitable ground near the sea-shore. For with this exception, the whole great country is a lonely wilderness, buried deep beneath massive glaciers and perpetual snows. Few mountain-peaks rise out of the vast bewildering sea of whiteness.

Yet even in that seemingly dead and ice-bound land God's forces are at work. Even in that awful waste of lifeless desolation there is perpetual change. Not alone in the fearful storms which sweep across the level expanse, filling the air with blinding snow. Apart from this, the change goes on. As in Swiss mountains on a small scale, so in Greenland on a gigantic scale, glaciers creep out from beneath the snow-masses, and make their way to the ocean, there sending masses of ice southward, to melt in warmer waters. And the winds from the south carrying vapours north, these vapours freeze and descend as snow or frozen mist upon the wide plains; thus keeping up the supply which through the glaciers is ever draining away. So the circulation of water, which goes on incessantly through all the world, goes on also in even those far-north regions.

The great Humboldt glacier of Greenland pours like other ice-rivers down its gorge, straight into the sea; and ends abruptly in a massive wall or cliff of ice, some sixty miles broad, and three hundred feet high. But this is its height above the water only. How deep it descends below the water is not known.

For these Greenland glaciers push their way out along the bed of the ocean, to a distance of several miles, into deeper and deeper water, holding toughly together, and resisting the buoying up tendency of the sea. The mass of the glacier cannot rise, since ice, though to a certain degree elastic, is unable to bend. When a certain depth is reached, the strain becomes too great for further resistance. At this point the strong upward pressure of the water causes huge masses of ice to split off and spring to the surface, making the ocean far around to seethe and foam like a cauldron of boiling water. These huge masses floating southward are called Icebergs.

In Switzerland the glaciers bear long trains of débris, dropped upon them from the crumbling peaks and cliffs.

Such *moraines* are far more rare in Greenland glaciers. The whole country is so completely buried beneath masses of snow, as to leave few peaks bare. Near the shore, indeed, the strewing of boulders on either side of the glacier increases; yet the great width of the glaciers makes it comparatively a small matter. Of all the icebergs which break away from the foot of such a glacier, only those coming from the extreme left or right would be likely to bear any large blocks. It is, however, not impossible that icebergs from the middle might have stones or rocks frozen in underneath them.

That many icebergs do carry heavy weights of material is undeniable. Captain Scoresby, who saw some five hundred icebergs in about 69° and 70° north latitude, describes many of them as bearing great loads of earth and rock, amounting in weight to fifty thousand or even one hundred thousand tons. Some of these bergs were a mile round; and many of

Greenland Glacier.

them most likely came from Spitzbergen, which, like Greenland, has its great glaciers.

Any rocks or stones carried off by icebergs are necessarily dropped upon the ocean-floor, when the iceberg melts.

The ocean-floor is not a mere flat plain, but consists, like the continents, of mountains, hills, table-lands, valleys, and lower levels. So it is easily understood how an occasional big boulder, brought from mountains far away, may be dropped upon a lofty peak or height beneath the ocean; which boulder, if that peak should ever be lifted up as dry land, might offer a curious appearance to geologists.

Such burdens on icebergs of the south Antartic Ocean seem to be even more common than on icebergs of the cold northern seas.

There is yet another mode by which stones and rocks are scattered over the ocean bed, and by which they may have been scattered, in ages past, over what was then the ocean bed and is now dry land.

Beneath the cliffs on Greenland shores, the surface of the sea freezes into a broad shelf or 'ice-foot.' This frozen platform is often one hundred and fifty feet broad, and rises as much as thirty feet above the water. It is attached to the cliff, and it follows every bend and curve in the coast-line. In the more northern latitudes it lasts all the year round, though varying in amount; while in the more southern parts of Greenland it breaks up and disappears every summer.

As the short summer, with its partial thaws, approaches, quantities of rocky rubbish fall from the cliffs upon this ice-shelf. Towards the north it is often buried beneath the collected piles of year after year, and is only relieved by the occasional breaking off and floating away of a large piece, when loose ice-floes are driven sharply against it by winds and currents.

But towards the south, as the warmth increases, the ice-foot is gradually demolished. Large portions heavily laden, are separated one after another, and borne away by the tide; each slowly thawing, and dropping to the bottom of the sea its pile of débris.

CHAPTER XXVII.

VOLCANOES.

'He looketh on the earth, and it trembleth; He toucheth the hills, and they smoke.'—PSA. civ. 32.

VOLCANOES are not scattered equally over all the earth. Here or there one such mountain may stand, or seem to stand, alone; but more commonly they are arranged in groups or lines. If, as many think, volcanoes are the outlets or safety-valves to vast underground reservoirs of fire, it is probable that one such reservoir often feeds several volcanoes.

A volcano is generally a cone-shaped mountain or hill, with a deep hollow or basin in the summit called a Crater. From this crater there are openings leading down into the underground fire-seas.

The cone of a volcano is usually built out of the materials thrown up from underground. Sometimes the cone is made of cinders, sometimes of tufa, a substance which comes from the hot cinders being wetted with heavy rain, sometimes of lava or melted rock; but more commonly of all three mingled. A new cone has been known to spring into being in a single day. Some volcanoes have only one cone and one crater, while others have several cones and several craters.

Volcanoes are divided into three classes—Active, Dormant, and Extinct. The active volcanoes are those which, from time to time, show signs of life by eruptions, more or less marked. Extinct or dead volcanoes are those which have so long remained quiet, that the fire-seas below are supposed to be exhausted. But this is seldom a matter of certainty. Volcanoes, thought to be dead, have suddenly proved themselves, by an unexpected outburst, to be alive, and to have been only dormant or sleeping.

Active volcanoes differ much in kind. Some are rarely known to be without signs of disturbance; others only have occasional outbursts, at more or less regular intervals of time. Some, during an eruption, throw out a large amount of solid material; others contain chiefly liquid lava. Some break out only through the crater. Others split open in any part of the mountain-side, and desolate the country round.

Volcanoes. 257

The tremendous nature of these underground fire-forces, and the enormous amount of molten rock which must be lying stored in earth's reservoirs, can be best known from such facts as follow.

In South America a long line of volcanoes stretches along the western coast, bordering the Pacific. Many of these Volcanoes of the Andes are believed to be extinct, but others continue active. Some of them outdo in loftiness Mont Blanc itself. Cotopaxi, for instance, is little less than nineteen thousand feet in height, being clothed commonly in a robe of spotless snow. This whole vast mass of snow has been known to vanish in a single night, under the tremendous heat of a sudden outbreak.

As a rule, not much lava is thrown from the volcanoes of the Andes, but more of vapour and ashes. Sometimes from outpourings of water, or sudden meltings of snow, tremendous rushes of mud have flowed down the mountain-sides and over the country round—the liquid mingling on its way with sand and stones. Valleys one thousand feet in width have been completely blocked up with such mud, to a height of six hundred feet.

In Mexico there are five great active volcanoes in a single chain, one of them being called Jorullo.

About a century and a half ago the table-land, from which the Jorullo cone now rises, was a fair landscape, where the indigo and the sugar-cane were cultivated, and the inhabitants lived in all seeming security.

Suddenly, in the month of June, 1759, suspicious underground rumblings and grumblings were heard, and by-and-by severe quakings of the earth followed. There were many who took alarm, and fled for safety. Well for them that they did so.

Two months of earthquakes were followed by the outbreak of flames from the ground, and masses of burning rock were flung high into the air. Then the whole surface of the country thereabouts seemed to be uplifted, like a huge swelling bubble, and vast quantities of lava mixed with cinders were poured forth, building no less than six separate cones. One who had lived there, and had tilled the land for many a year, watched from a distant height this strange transformation of his peaceful farmstead into a fiery furnace.

Forty years later the great upheaved mass, with its smoking cones, was still warm. Two little rivers, which had once flowed in the fair plain, had vanished altogether, and were heard of no more.

Some of the mightiest eruptions known, so far as

VOLCANO.

regards the amount of lava poured out, have been those of Iceland.

The year 1783 was remarkable for the great outbreak of the Icelandic Volcano, Skaptár Jokul. The first sign of coming mischief was the bursting out of a volcano under the sea, about thirty miles from land. A new island was built up out of the materials there belched forth from beneath ocean's floor, and the King of Denmark claimed it for his own, naming it 'New Island.' His Majesty enjoyed but a brief possession. One year went by, and the loose pile had been washed away by the restless waves.

Meanwhile earthquakes became more and more severe in Iceland, and at length the threatened eruption came.

A fierce torrent of liquid lava burst from the crater of the Skaptár Jokul, and poured down the mountain. The river Skapta flowed below in a gorge between steep rocks, two hundred feet apart, and from four to six hundred feet high. The lava took the course of this river-bed, drying up the stream, filling up the whole gorge, and pouring over the lofty cliffs into the fields on either side. Still pressing onward, it reached the end of the rocky gorge, where a deep lake used to stand. The lava invaded the lake, banished the water, filled up the entire hollow, and again advanced.

After a while it reached a mighty cataract, where anew it took the place of the water, pouring over in a stream of liquid rock. Thence it spread widely over the lower countries, carrying desolation wherever it went.

By this time the Skapta channel was completely blocked, and still the lava stream poured unceasingly from underground. It now took a new course, started in another direction, invaded a second river, filled another deep gorge, and spread itself out again over another part of the lower country.

Of these two streams, composed of fiery melted rock, one was fifty miles in length, the other only five miles less; one spread itself out to a breadth of fifteen miles at its widest, the other to seven miles; while each was for a considerable part of its course as much as a hundred feet deep, and in the rocky defiles no less than six hundred feet. Some twenty villages were destroyed, and about nine thousand people lost their lives.

It has been reckoned that the amount of lava poured out from Skaptá Jokul in those few months, was sufficient to make an entire mountain as large as Mont Blanc.

The volcano Etna, rising to a height of nearly

eleven thousand feet, may almost be described as a mass of volcanoes, rather than as one. It has indeed a chief cone, and a principal crater, but it has also two hundred or more lesser cones with their lesser craters, outgrowths from itself. These spring from and circle round upon the mighty central cone, somewhat after the fashion—to use a rather inappropriate simile—of a hen-and-chicken daisy. One of these lesser cones is described as seven hundred feet high. Some of them continue still to smoke, while others are overgrown by trees. From the great centre crater sulphureous vapours are perpetually poured forth, and thence from time to time come streams of lava.

A great Etna eruption took place in 1669, and one of the lesser cones, Monte Rossi, was then formed.

There was first a warning earthquake, which levelled a whole town in the neighbourhood. Next, in a plain near, a tremendous ground-crack suddenly appeared, splitting to a distance of twelve miles. It was about six feet wide, and shone with a lurid light, from the glowing lava within. Five more such huge cracks opened alongside.

Lava then poured out in a stream from the new cone, Monte Rossi, at that time formed or being formed, and as it poured it rapidly overwhelmed fourteen towns and villages.

The inhabitants of Catania, in dread of such an event, had surrounded their town with a strong wall, sixty feet high. The flood of liquid rock streaming over the country reached Catania, piled itself slowly higher and higher, till the top of the wall was reached, and then flowed over, deluging the nearer part of the town. The wall remained standing, and to this day the cold lava may be seen, looking as if petrified in the act of creeping over the rampart.

This lava stream journeyed its first thirteen miles in twenty days, but cooling steadily as it moved, it took twenty-three days for the last two miles. When finally it reached the sea, it was still forty feet deep and nine hundred feet wide.

A fearful outburst took place in the island of Java many years ago.

The mountain Galongoon, up to the year 1822, showed no signs of disturbance, being clothed in a thick growth of forest. The country around was cultivated and peopled. At the top of the mountain there might indeed be seen a cup-shaped hollow, but no records had been handed down of any former outburst, and no expectations of evil were felt.

All at once in the month of July, the waters of the Kunir, a river close at hand, became heated, with no apparent reason. Nearly three months passed without further tokens of mischief. Then, on the 8th of October, a tremendous explosion was heard, and the ground shook beneath men's feet; while hot water and boiling mud, with brimstone and ashes, were poured upwards out of the mountain-top, like a huge ascending water-spout, which rose high before it fell to earth and deluged the country.

So tremendous was the force exerted, that some of the matter thrown out in this jet reached the ground at a distance of forty miles from Galongoon. For a distance of twenty-four miles in that same direction, the land was fairly inundated by bluish mud, to such a depth that villages were entirely buried beneath it.

The boiling mud and red-hot cinders were flung out with such violence, that they passed in a great measure over the nearer villages, and did most damage to those lying farther away.

Many human bodies literally boiled in mud were strewn about. The first outbreak went on for about five hours, and was followed by heavy rain. Four days later another yet more fearful outburst took place. Again hot water and mud were poured forth, and huge basaltic rocks were flung bodily to a distance of seven miles, as a child may toss a pebble across the road; while a great earthquake shook the island, and one whole side of the mountain broke down, an immense gulf being thus suddenly formed.

Truly we find a lesson here for the student of Geology, as to the mode in which great changes on the earth's surface may here or there, at one time or another, have been rapidly brought about.

It was said that over one hundred villages were destroyed, and that over four thousand people were killed, in this eruption.

Two good opposite instances of volcanoes containing chiefly solid and chiefly liquid kinds of matter, are the volcano of Vesuvius in Naples, and the volcanoes of Hawaii in the Sandwich Islands.

Vesuvius is the chief of a volcanic group. The island of Ischia, belonging to this group, was greatly troubled by earthquakes and fiery outbreaks, before the Christian era, in times when Vesuvius was looked upon as an extinct volcano. From the time when the fires of Vesuvius began to play, Ischia enjoyed quietness up to the year 1302. An outbreak then took place, and again it lived in peace up to the present year. While this chapter is being actually written, another outbreak has occurred—March, 1881—and at least two hundred people have been suddenly and awfully cut off.

Some think that there may be a connection between the underground fire-seas of Vesuvius and Etna. It has been noticed that when one great mountain is active, the other appears usually to be at rest. If a single vast reservoir of liquid lava and furnace-heat lies below the two, reaching from one to the other, we can easily understand how both safety-valves would not be in action at the same time.

In early ages Vesuvius was looked upon as an extinct or at least as a dormant volcano. The first known eruption was the famous one of 79, wherein the towns of Herculaneum and Pompeii were buried beneath showers of ashes.

A great eruption took place in the same volcanic neighbourhood, in the year 1538. Earthquakes shook the country; fire burst from the ground; ashes, stones, and water were poured forth; the sea was driven back from the Bay of Baiæ; the solid ground was uplifted in the form of a huge bubble: a mouth or crater opened in this bubble, to pour out stones, ashes, and mud; and in less than a week—chiefly in the course of twenty-four hours—the Monte Nuovo or New Hill was formed, being over four hundred feet in height, and a mile and a half round at its bottom.

A description of the Vesuvian eruption of 1779, given by an eyewitness, says that—'Jets of liquid lava, mixed with stones and scoriæ, were thrown up to the height of at least ten thousand feet, having the appearance of a column of fire.' All this matter, falling back upon the cone and shining brilliantly with a 'lurid red light, seemed to be one vast mass of fire, sending heat to a distance of six miles around.' In another such eruption 'millions of red-hot stones were shot into the air, full half the height of the cone itself, and then, bending, fell all round in a fine arch.'

In the early part of the present century, the great crater of Vesuvius had been slowly filled up with lava rising from below, or with other materials tossed up in lesser outbreaks. The crater was, in fact, scarcely a *cup* any longer, or at least it was no empty cup. When the eruption of 1822 took place, all these collected materials were flung clean out in one mighty effort, and once more a great empty hollow was left, three-quarters of a mile across. So strong was the explosion which worked this sudden clearance, that about eight hundred feet of the mountain-top were blown completely away by it.

Although lavas flow from Vesuvius, yet a considerable proportion of the material thrown up is of a more solid nature such as granite, sand, stones, cinders, and dust.

A marked difference is seen in the volcanoes of the Sandwich Islands. The chief volcanoes of Hawaii, Mount Loa, Mount Koa or Kea, and Mount Kiláuea, are more or less mere shells, filled with very liquid lava.

Eruptions in these mountains commonly take place, either through the breaking down of part of the crater-brim, from the weight of the rising liquid within; or through the opening of a sudden crack or rent in the side of the heavily-charged mountain. Either way the lava streams down and inundates the country.

The following extract so well bears upon the subject of Geology that I cannot forbear quoting it:

'At Hilo . . . they have felt the perpetual shudder of earthquakes... Once they traced a river of lava burrowing its way fifteen hundred feet below the surface, and saw it emerge, and fall hissing into the ocean. Once from their highest mountain a pillar of fire, two hundred feet in diameter, lifted itself for three weeks one thousand feet into the air, making night day for one hundred miles round, and leaving as its monument a cone one mile in circumference. *We* see a clothed and finished earth; *they* see the building of an island, layer on layer, hill on hill, the naked and deformed product of the melting, forging and welding, which go on perpetually in the crater of Kilauea.'*

* 'Travels in the Sandwich Islands,' by I. Bird.

And again, with reference to Mount Loa: 'It is probable that the whole interior of this huge dome is fluid; for the eruptions from this summit-crater do not proceed from its filling up and running over, but from the mountain-sides being unable to bear the enormous pressure, when they give way, high or low, and bursting allow the fiery contents to escape. So in 1855 the mountain-side split open, and the lava gushed forth thirteen months, in a stream which ran for sixty miles and flooded Hawaii for three hundred square miles.'

In the summit of Kilauea there are open lakes of liquid fiery rock, described by travellers as fearfully sublime and beautiful. You have seen the bubbles which break out upon the surface of water boiling in a pot. Such bubbles are seen upon the great lava-lake of Kilauea—a boiling pot one thousand feet across, the bubbles being fire-fountains, thirty or forty feet in height, playing majestically over the glowing surface.

In these Hawaiian outbreaks of lava, earthquakes were not commonly known to take place; but in 1865 there came an eruption of exceptional nature. So fearful and continued were the quakings of the solid ground beforehand, that men held their breath for fear. Houses fell shattered; trees rolled to and fro, slashing the air; people sat clinging to the earth, rocked helplessly from side to side; the ground gaped in thousands of places; and the whole country 'quivered like the lid of a boiling pot.' In one place three hundred shocks were counted in a single day; while in other places they were uncountable.

Then appeared suddenly great rents in the mountain-side, and lava-jets shot madly upward to the height of a thousand feet. Rivers of lava poured seaward from these fissures, turning a fair country into a scorched wilderness, wrecking villages, destroying life, making havoc of all that lay in their path.

During more than a week four distinct jets or fountains continued to pour upward out of the rents to a height of five hundred or a thousand feet. At the same time the crater of Mount Loa, and also the crater of Kilauea—the latter being twenty miles distant,—which before the eruption had been filled high with liquid lava, were gradually emptied.

CHAPTER XXVIII.

EARTHQUAKES.

'The Lord hath His way in the whirlwind and in the storm
. . . . the mountains quake at Him . . . and the rocks are
thrown down by Him.'—NAHUM i. 3, 5, 6.

THOUGH Earthquakes are often a mere accompaniment to volcanic outbursts, taking place in volcanic districts; yet they often happen also in countries far removed from volcanoes, with no seeming connection between the two. There can be no doubt, however, that the actual cause of earthquakes is connected with the cause of volcanic eruptions.

A few particulars will now be given, more especially in reference to earthquakes; those in the last chapter having been more especially in reference to volcanoes. The object in bringing them forward is still the same,— to show the workings of the great underground agent, Fire; and to draw attention to the extreme uncertainties which exist as to the manner and the speed of earth-crust formation in the past.

Among many severe earthquakes which have been known to take place in New Zealand, there was one in 1855 which completely altered the appearance of the coast for a considerable distance. One small cove was described as having been, in a single night, changed into dry ground. The extent of land and water affected by this earthquake was three times as much as the whole of the British Isles.

In another New Zealand earthquake, of a few years earlier, a great rent or 'fault' was caused in the mountain-strata. This split and slip, which took place *suddenly*, was only about eighteen inches wide, but it ran through the rocks for a distance of sixty miles. A hint lies here for us, as to how the greatest 'faults' in America and elsewhere may have been produced.

In 1835 a severe earthquake was felt in Chili, through about one thousand miles of country, running north and south.

Some years earlier there had been one in the same country, far more destructive. Through many months shocks went on almost continually, the most violent shock being on the 19th of November, 1822. It reached through a distance of twelve hundred miles; and next morning men found that the entire coast had been bodily uplifted to a height of two, three, and four feet, varying in different parts, while inland the sudden upheaval must have been as much as six or seven feet. It was calculated that about one hundred thousand square miles had been thus raised in one tremendous effort of nature, remaining afterwards at its new level.

Another instance, somewhat like in kind, happened near Cutch, in India. A violent earthquake took place there, not far from the beginning of the present century, doing much damage. An estuary of the sea, where at high tide the water had been six feet deep, and at low water only one foot, became suddenly eighteen feet deep at low water; while the neighbouring fort and village of Sundree were overflowed by the ocean.

Some two thousand square miles of land were then and there transformed into an inland arm of the sea. Also, immediately after the shock, the inhabitants of Sundree, watching from a spot where they had fled for safety, could perceive a long raised mound, about five and a half miles distant, where before there had been a low flat plain. This mound, over fifty miles long, sixteen miles wide, and about ten feet deep at its most, was named by them "Ullah Bund," or "The Mound of God." A strange *folding* of the earth-crust seems there to have happened; sharp rising and sinking side by side.

In 1812 fearful earthquake-shocks were experienced in Caraccas, South America. The ground rapidly rose and fell, with terrible sounds beneath, and in a single moment the whole city became one vast pile of ruins, with ten thousand human beings buried in the wreck. Lava and water were thrown from a volcano not far distant.

Somewhat before this event, and possibly connected with it, great disturbances took place in South Carolina and Missouri—one of the comparatively rare instances of severe and long-continued earth-shaking in places far distant from any volcano.

Tremendous changes are described as having come about with awful suddenness. Land became covered for many miles with water, and then became dry land again. Lakes, twenty miles across, were formed in the course of a single hour, and others were as rapidly emptied of all their contents. The New Madrid graveyard was launched bodily into the Mississippi; and the inhabitants of the town told afterwards how the ground had risen and sunk in great billows like the sea; and how, when these solid billows reached a certain height, they broke open, and water with sand and coal were spouted out to the height of the tree-tops. Hundreds of these gaping cracks were seen seven years afterwards by a traveller, remaining still unclosed.

The shocks continued through three months, and the people gradually found that the cracks or fissures usually opened in a particular direction; so that, by felling large trees to lie in an opposite direction and taking refuge on the trunks, they sometimes escaped the fearful death of being swallowed alive, as were many of their number, by the opening earth.

Another great earthquake, in many respects similar, happened in Calabria towards the close of the last century. *An* earthquake it can hardly be called, for though the shocks began in February, 1783, they went on during four years. The ground often swayed and heaved like the surface of the ocean, the motion being sometimes so strong that trees were seen to bend and touch their tips to the very earth, like an Eastern making his salaam, righting themselves again as the vibration passed on.

In some parts the ground rose, in others it sank. Deep fissures were formed, and remained open. Also, as the earthquake-wave swept by, many cracks yawned suddenly, without sign of warning, and swallowed men and beasts alive, the walls of the rent closing quickly upon them. In some rare instances, it was said that when people were thus swallowed and buried alive, another earthquake-wave following immediately, the same cracks opened again and flung out their living victims, with accompanying jets of water. A fearful experience truly to have lived through!

In one place the cracks or fissures, instead of being regularly placed, ran branching every way from a centre, like the lines on a starred pane of glass. These remained permanently. One fissure, in another part, which after the earthquake was merely a big crack about a foot in width, had yawned so broadly as to swallow an ox and almost one hundred live goats.

About forty thousand people were believed to have lost their lives directly through the earthquake-shocks, and about twenty thousand more indirectly, through sicknesses caused by the earthquake.

The famous Earthquake of Lisbon in 1755 is too well known to require close description, yet I can hardly pass it entirely over.

It came with terrible suddenness. A sound like

underground thunder was heard, a tremendous shock followed, and within six minutes about sixty thousand people were destroyed. The sea drew back, then rolled tempestuously in upon the land, as a wave fifty feet higher than its usual level, sweeping all before it. Neighbouring mountains were shaken and rent, flames being seen to spring from them, and large masses of rock were flung down into valleys.

The newly-built quay of Lisbon, upon which people had flocked for safety, sank suddenly down, and vanished into some unknown abyss. Not a man standing on it was ever seen again; and the water, which in that part had been only thirty feet deep, was said to have gained all at once a depth of six hundred feet.

This earthquake was felt over an enormous distance; the whole extent of land and sea affected being at least four times the size of Europe.

The thrill reached to the Alps, to Sweden, to Germany, to Great Britain, to the West Indies, to the Canadian Lakes, to the north of Africa. About eight miles from Morocco, the shock was so violent that a whole village was swallowed bodily at one huge gulp, the earth opening and closing upon the buildings with all their eight or ten thousand inhabitants. The

Earthquake at Lisbon.

same shock, extending through the ocean, sent great waves upon the land in many different places, as at Cadiz, at Tangier, and at Kinsale.

An earthquake which took place in Jamaica, nearly two hundred years ago, should perhaps be mentioned. Here as elsewhere the ground heaved and swayed like a stormy sea, and burst into countless rents—two or three hundred such cracks being often seen at once, opening and closing, as the earthquake-wave passed on. Many individuals were swallowed alive in these earth gashes. Some, as in Calabria, were buried for an instant, and then flung out again. Others were caught by the middle, and were squeezed to death, as the gaping jaws of the chasm shut upon them. Others thus seized, had only their heads remaining above ground.

Something may be gathered from these particulars as to the work done by earthquakes in the fashioning of the earth's crust. Mountains have been shattered and split, cracked and faulted. Hills have been formed in a single night. Forests have been levelled at a blow. Miles of country have been suddenly lowered or suddenly raised. Valleys, ravines, fissures, have been instantaneously formed or deepened and widened. Sea-beaches have been lifted or depressed; water-coves have become dry land; lakes have been made or emptied; all in the course of a few hours. Towns and villages have been laid low, wrecked, buried underground, or engulphed in the ocean, with scarcely a moment's warning.

With regard to volcanic eruptions, it is calculated that, taking large and small together, there may be about twenty in the world every year, on an average, or two thousand every century.

Even supposing that there have never been any mightier or more frequent outbursts than in modern times—a question about which we are necessarily in the dark,—the amount of change worked in the earth's crust, by two thousand volcanic outbursts each century through countless ages, must indeed be great.

Where actual eruptions have taken place in the past, signs of the same are often still visible in the shape of cones or lava. In Auvergne, for example, there are many such cones, once fiery and active, now cold and dead.

But in other parts, where no such silent witnesses are found, we are not thereby freed from uncertainties. Rather, we are plunged more deeply into them.

For if we may say with some confidence that volcanoes have not existed here or there, no such assertion can be made with regard to earthquakes. These abrupt movements of the crust reach to unknown distances from volcanoes, and their effects cannot in after-ages be distinguished from the effects of quieter alterations slowly taking place.

There is not a country in the world which may not, at one time or another, have endured some of these terrific shakings. There is not a spot in the earth which may not have been upheaved or lowered, rent or dislocated, by earthquake-action. There is not a stratum in the earth-crust building which may not have been more or less affected by these underground influences. There is not a fault or a slip or a slide in the rocks, there is not a displacement or a rise or a fall in the strata, there is not a bend or a twist or a fold in the layers, which *may* not have been the sudden and rapid result of disturbances below. Every chasm, every valley, every table-land, every mountain, which may be the result of slow and gradual water-working, *may* no less be, at least in part, the result of sudden and tremendous fire-working. The uncertainty in which we stand on such points, should warn us to be careful in drawing conclusions. It has been said:

'Give me an earthquake, and I will give you any physical condition you please.' We can scarcely make too much allowance for these past unknown possibilities.

CHAPTER XXIX.

HOT SPRINGS.

'Worship Him that made Heaven and Earth, and the Sea, and the Fountains of Waters.'—REV. xiv. 7.

A FEW more particulars still have to be given, as to the underground fiery forces, of which the volcano and the earthquake tell us so much.

The risings, sinkings, and tremblings of land, detailed in the last chapter, although often reaching to a very great distance from any volcanic centre, were yet in almost every instance plainly connected with one or another such volcanic centre.

But movements of the earth-crust do also take place, which cannot be distinctly traced as taking their rise at any such centre: albeit there is little doubt that they spring from the same cause.

The examples given have been of rapid movement and sudden change. The underground forces do not however, always work either rapidly or suddenly. There are gradual risings and gradual sinkings, as well as sudden and startling upheavals and subsidences.

In the beginning of the eighteenth century, signs were first observed of a certain change taking place on the coasts of lands bordering the Baltic Sea and German Ocean. Rocks once buried under the sea had become visible at low tide; towns which once bordered the sea had become inland cities; and former islands had become part of the mainland. Nay, ancient history had spoken of the whole of Scandinavia as an island, whereas it was then distinctly joined to Europe.

Plainly, therefore, men argued, the sea was beating a retreat. It was evident that the waters of the Baltic Ocean and the German Sea were gradually sinking.

This idea roused opposition, and no wonder. For if the water were sinking lower in those two seas, it must have been sinking lower all through the Atlantic Ocean; and if throughout the Atlantic, then throughout the Pacific Ocean also, and in fact all over the world wherever open sea existed. If the whole ocean had really sunk at the rate calculated—some forty Swedish inches in one hundred years, or as much as fifteen feet during four hundred years—why were not the same changes seen along all sea-coasts in all the world? How was it that the change appeared, even in Sweden, more marked at one spot than at another? Near Stockholm the waters seem to have sunk ten inches in a century, while at some distance to the north of Stockholm the alteration was two feet and a half in a century. If the sea were sinking at all, it must surely sink equally everywhere.

So the idea of the sinking sea was given up, and the only other possible explanation was that the land had slowly risen. This is held to be the true explanation.

The rise is not the same in all parts, but it is everywhere very slow and steady. How long it has gone on, or will continue to go on, we cannot tell.

The remarkable part of the matter is, that—so far back at least as history reaches—no volcanic outbursts or earthquakes have happened in Sweden. A slight tremor may indeed have thrilled the land from some distant disturbance, as when the great Lisbon earthquake vibrated through Europe. But Sweden is no volcanic centre, and shows signs of no underground fire-seas. Yet doubtless this gradual rising is in some manner connected with the underground fiery worker.

Greenland is, in like manner, a land peculiarly free from volcanic heavings or shakings; yet in Greenland also a somewhat similar change is taking place. The difference is that while Sweden is rising, Greenland is sinking.

For more than six hundred years past the coast of Greenland has been slowly going down, and the waves have been gradually creeping higher. The Greenlander is much too wise to build his hut close to the sea. In one place there are strong poles still visible under water, to which once upon a time the Moravian settlers used to fasten their boats. They had to retreat inland and leave their poles behind them.

It will be seen in the next chapter that a large part of the floor of the Pacific Ocean is also thought to have been long gradually sinking, though about this we cannot be sure.

Many risings and sinkings of land in past ages were spoken about in the second part of this book. The examples given in the present chapter and the one before, will show clearly how such sinkings may have come about either suddenly or slowly—either as some great catastrophe of an hour, or as a quiet change lasting through centuries.

One remarkable instance of these variations in the

earth's surface is to be seen in the ruined Temple of Jupiter Serapis, at Puzzuoli, not far from the Monte Nuovo.

Three pillars remain standing, each one about forty feet in height, Above the pedestal of each rise twelve feet of smooth uninjured marble, and over them are nine feet, where borings through and through the marble have been made by a certain ocean shell-creature.

This seems to show clearly that at one time the temple must have sunk so low—through the sinking of the ground—that the twelve feet of smooth marble were covered up and protected by earth or rock, while the *nine* feet must have had ocean water flowing round them. Again an upheaval must have taken place later, lifting the ruined temple with its three standing pillars to their present position. The temple is believed to have been built long before the Christian Era. The downward and upward movements in this case were probably very slow. Had they been otherwise, the three pillars could scarcely have remained upright.

By far the greater number of volcanoes in the world are placed at no great distance from the ocean; and it is supposed that water may have much to do with their eruptions.

The most commonly received explanation of volcanoes is that of vast underground fire-seas or fire-reservoirs, connected with the cone-shaped hills above ground by natural openings. As already explained, the cone-shaped hills are usually made entirely in the first instance out of materials poured up from below. A volcano may be looked upon as a necessary safety-valve to such a buried fiery furnace of tremendous heat and melted rock.

Now it is believed that much water from the sea soaks into the nearest land, finding its way through the loose sand or through cracks and crevices in the rock. Thus it wanders on till it joins other streams, and at length finds its way up to the surface.

But where these fiery lakes lie hard by, a different result is likely to follow. Imagine the enormous extent of the reservoir which, for instance, supplies all the volcanoes along the mountain range of the Andes, or that which extends beneath the great Hawaiian group, or that which feeds the mighty Vesuvian neighbourhood. If sea-water in any large quantity should soak through the soils, and find its way to these fiercely-glowing reservoirs, we can imagine the tumult which must ensue. Every drop of water would be rapidly turned to steam, and the effects of large bodies of steam suddenly formed in a limited space are well known. Steam has mighty explosive power, as seen in numerous fearful accidents of bursting boilers and lost lives. This explanation may account for many great shakings and tremblings of land.

The same, modified, would also serve in some degree for the Hot-water springs of some countries.

Such springs are found in many places, and often far removed from known volcanic centres. Look at the hot springs of Bath, for example. Bath is built in a basin surrounded by hills, probably the crater of an extinct volcano. No outbursts have been known to take place there within the memory of man, and no signs remain of outbursts in earlier ages.

Yet in the bottom of that hill-encircled hollow, four streams of hot water, laden with mineral torn from the rocks, rush perpetually up from underground, and have so rushed for centuries past. One alone of these springs sends forth eight gallons and a half every minute. The heat of the water is from about $114°$ to $120°$ Fahrenheit. The amount of mineral borne upwards in a single year by all four springs, has been calculated to be large enough for the making of a square column, one hundred and forty feet high and nine feet in diameter. Heated gases, to a large amount, are poured out with the water.

The nearest known volcano lies four hundred miles away. But what if, below the pleasant town of Bath, there still lie the smouldering remains, deep underground, of the fiery reservoir from which the crater was probably once fed? Water passing through the rocks in channels might thus be heated, and the temperature of the springs might thus be explained.

Many more such hot springs are found in different parts of Europe and America.

A still more remarkable description of hot spring is the kind called a Geyser. The Geysers are, in fact, natural hot fountains, playing at intervals, with pauses between. There are Geysers in Iceland, in New Zealand, and in North America.

About thirty miles from the Icelandic volcano, Hecla—with which no doubt they are connected—one hundred Geysers may be found, within a compass of two miles. They break out from thick layers of lava, through which the hot streams have forced their way. The water rises through a natural pipe, from underground regions, into a cup-shaped basin, larger or smaller according to the size of the particular Geyser.

The great Geyser has a basin fifty-six feet in diameter one way, and forty-six feet the other. A

GEYSERS.

mound surrounds it, built out of the flinty droppings from the waters. The play of the stream is up a pipe some ten feet in diameter, and known to descend abruptly seventy-eight feet. Sometimes the basin is empty, but more commonly it is full of clear boiling water. The eruption, as a rule, comes on gradually, with rumbling underground noises and shakings of the earth. The water is flung up in jets, with loud explosions, becoming more and more powerful, until the jets reach a height of one hundred and fifty or even two hundred feet. Clouds of vapour float away, and at length the flow of water stops. A sharp rush of steam up the pipe, with a thundering noise, closes the display.

Most of the Geysers play for about five or six minutes at a time, though sometimes they will go on for half an hour. It was found that, by throwing large stones down the pipe of one Geyser, an eruption could at any time be brought on. The stones, exploding into pieces, were thrown violently to an unusual height.

The Geysers of Iceland are far surpassed in numbers by those of America. In the Yellowstone Park district there are hot and warm springs, together with Geysers, amounting in all to some ten thousand, already known, while an unexplored region of them lies beyond.

The 'Giant Geyser' in this neighbourhood has a partly broken-down cone, ten feet high, and twenty-four feet in diameter close to the ground. Its occasional hot jet of water rises to a height of one or two hundred feet.

The 'Beehive Geyser' throws out a jet to the same height, though its basin is very much smaller.

Another, called 'Liberty Cap,' has quite ceased to play, and is supposed to be extinct. While the 'Beehive' cone is only three feet high, that of 'Liberty Cap' is thirty feet.

'Old Faithful' is yet another, so named because of its curious regularity in action. Once in every sixty-five minutes, as a rule, it flings a jet to the height of a hundred and thirty feet.

In one instance, two Geysers were seen to play alternately a duet of jets; one ceasing immediately the other began, and beginning as soon as the other left off.

CHAPTER XXX.

CORAL.

'Thou hast made. . . . the seas and all that is therein.'—
NEH. ix. 6.

FROM liquid lava and boiling water, from fiery outbursts and fearful earth-quakings, we turn now to quite another class of workers.

For all lands in the world are not built up by rivers or piled together by volcanic eruptions. There are lands—not indeed so wide in extent—quietly raised, inch upon inch, through century after century, by the ceaseless activities of the soft jelly-bodied Polyp of southern seas, called the Coral.

Coral-animals lived once upon a time over England, when half-built English shores lay low under the ocean-waves, and over many other countries also of the temperate zones, where in these days they cannot exist. To learn about the coral now, we must wend our way to the warm soft clime of the South Pacific Ocean or the Caribbean Seas.

The hard substance, commonly white in colour, which we call coral, is made chiefly of lime, and is in fact a sort of inside skeleton to the soft-bodied living animal, the coral-polyp. When the coral-reef is growing, the slimy body of the jelly-like polyp is spread over the outside of the hard coral, busily gathering lime from the ocean-waters, and forming more and more of the hard substance, which lasts long ages after the delicate living creature has died. But if you look at a piece of coral, you will see it to be full of tiny holes or cells, many of them so small as to be like mere pin-pricks. Into these holes the polyp can almost entirely withdraw itself if alarmed. So the hard coral is only in part a kind of inside skeleton, since it serves also for an outside protecting shell.

There are many different kinds of corals, and each kind thrives at its own particular depth in the ocean. Some descriptions are found as low as six or eight hundred, or even nine hundred feet, below the surface; but these are not reef-building corals. They are usually a solitary description, either living quite alone, or else living just a few together, in which case the coral formed by the little company of polyps is generally branched.

The reef-building kinds are not 'deep-sea corals,' but are found, as a rule, never to exist at a greater depth than one hundred and twenty feet. Also, neither they nor any other coral-polyps can live above water. So the work of reef and island-building has all to be carried on—so it appears to us—within the belt of water reaching from low-tide level to one hundred and twenty feet downwards. When such coral is found either below that depth or above water, it is dead—the hard white skeleton-substance remaining, with its little empty holes, and no living polyp.

If this be so, how is it that coral lies above reach of the waves? Also, how can coral be found—as it certainly is—far deeper down than one hundred and twenty feet?

The first of these questions is not difficult to answer.

For there are other powers at work beside the busy animals. The coral polyps carry on their formation steadily, inch by inch, till they have built a broad platform up to the low-tide surface of the sea. There they stop, for they have no power to make farther advance. They may lengthen or widen the bank of coral; they cannot raise it higher. If the waves should cease to wash over them they would die.

But the sea carries on the unfinished work. The heavy swelling surge of the broad Pacific beats ceaselessly against and over the bank, and the waves break off masses of coral, flinging them on the platform and heaping them together, till at length a height is gained over which only the stormiest spring-tides can sweep.

You must not suppose that the whole reef is composed of delicate coral branches, such as you have seen in shops. This reef-coral is a very firm and solid kind, not red or pink, but white; and the perpetual grinding of the waves wears vast quantities of it into fine powder. The greater part of the reef below is composed of hard limestone, made out of the powdered coral; while over the surface, as soon as that surface is raised high enough, lies a thick layer of the same white powder—the bright white sand of the coral-island beech. Our yellow sand is flinty in nature, but the sand of a coral-island is made chiefly of lime.

Mingled with broken and ground-up coral of the reef there are great quantities of shells, small and large; lower down bound together into firm rock, higher up loose and mixed together, more or less broken and pounded.

When this stage in the reef-making is reached, the next step is that seeds of plants and trees are carried thither by the waves, and find a resting-place upon the little ledge. These spring into life, and grow quickly in the tropical climate. Now and then whole tree-trunks are borne to the island, having on them insects or lizards swept from some distant shore; and so life begins there. Sea-birds, too, settle from time to time; and land-birds, driven by gales from their native homes, take refuge in the slender belt of young trees soon growing along the reef.

The coral-buildings are of different forms. Sometimes they are found as islands, and sometimes as long narrow reefs. The islands vary much in shape, but the commonest and also the most remarkable kind is the Atoll.

An Atoll is simply a *ring* of land—or rather of coral—surrounded by the deep ocean, with a lake or 'lagoon' of shallow salt water in the middle. Upon this ring of land, with its inner and outer beaches of pure white sand, tall cocoanut-trees grow abundantly. Inside, the water is pure and still and clear, often of a vivid green colour. Outside, it stretches on every side to the horizon, profoundly blue; while around the slender circular strip of coral the fierce Pacific surge thunders unceasingly, breaking down, and tearing up, and grinding to powder, the materials of which the island is composed. And hour by hour the soft transparent polyps are at work, gathering fresh lime from the foaming breakers, closing breaches, repairing damages, and saving the tiny ocean-oasis from destruction.

The coral-polyps certainly give a good illustration of the advantages of combined labour. Weak as they are individually, they are more than a match, united, for the mighty waves. Perhaps it is a little doubtful how far we may fairly speak of the reef-building coral-polyps as 'individuals.' They are bound so closely together in their life and labour, that if one takes in food, he nourishes his neighbours as well as himself.

A certain traveller, who visited many of these islands in the Pacific Ocean, found that out of thirty-two, as many as twenty-nine had lagoons in their centres. The largest lagoon was thirty miles across, the smallest less than one mile.

All these islands were composed of living and growing coral except one, and that one was singularly unlike the rest. It had no lagoon, but was about five miles long by one mile broad, having a flat surface, and upright cliffs all round of dead coral, fifty feet in

CORAL ISLAND.

height. This island seemed to have been forced up to its present level by some great underground thrust—doubtless the result of volcanic forces. In another island the lagoon had apparently disappeared through the building up of coral all over it.

In the atolls the ring of land has always at least one opening, through which ships may pass into the lagoon, and there find a safe harbour. The reason for there being this opening is not yet quite clearly understood, though its convenience for sailors is plain enough. It is usually found to the windward of the island. The most probable explanation seems to be the need for some outlet for the fresh-water drainage of the island, resulting from rain. Wherever that outlet might lie, coral-building would be checked, since coral-polyps have a strong aversion to fresh water.

There are, in the Pacific, groups of coral islands which extend over hundreds of miles. But the islands composing such groups are usually far scattered, and for the most part small in size.

The Maldive Islands stretch through four hundred and seventy miles in one direction, the long chain having a rough breadth of fifty miles. A remarkable point about these islands is that most of them are not simple atolls, but *atolls of atolls*. A child's chain of dandelion stalks would best illustrate this. Each link is one small ring, and all the small rings joined together form one large ring.

Some of these complex islands are from forty to ninety miles in diameter. The centre is occupied by the great chief lagoon, its clear waters varying in depth from about fifteen to fifty fathoms. But the ring of land surrounding, instead of being a plain broad reef, is a string of little rings or atolls, some of them from three to five miles across, and each having its own tiny lagoon. Occasionally a few more such tiny atolls are scattered about in the large central lagoon. Outside the ring of little atolls the ocean-waters become suddenly so deep as to be almost unfathomable.

In addition to atolls and other islands, the polyps often build long *reefs* of coral. These are sometimes called Fringing-reefs, and sometimes Barrier-reefs.

The fringing-reefs are so close inland as to be joined to the shore. The barrier-reefs lie farther out to sea. They usually border an island, or run along the coast of a continent. Such reefs are often more or less wooded like the coral-islands, and sometimes they are very extensive. Off the Feejees there are huge barrier-reefs from five to fifteen miles wide. Near

New Caledonia there are reef-grounds four hundred miles long. Beside Australian coasts there are barrier-reefs, fifty miles away from the shore, lasting with breaks for a distance of one thousand miles. The coral structure in these reefs descends to a depth of thousands of feet.

But how can this be? What about the fact above-stated that the reef-coral polyp cannot live below one hundred and twenty feet of water-depth?

Many theories have been put forward in explanation. At one time it was thought that the circular form of the atolls was probably caused by their being built upon the edge of a volcanic crater under the ocean, the said crater, filled with water, forming the shallow lagoon.

This theory is not now so widely held. The present and more generally accepted idea is that of a gradual sinking of land—or rather of the sea-bottom—throughout a great part of the Pacific Ocean.

That the ocean-bottom does so sink, or has so sunk, slowly and quietly through ages past, is a matter about which we have no direct proof; but it *may* have been thus in the Pacific, as in Sweden and elsewhere. This theory explains the mystery better than any other yet offered.

Suppose that in the deep Pacific Ocean a certain

mountain once lifted its head, as a small island, above the water. All islands in mid-ocean are in reality hill-tops or mountain-summits.

In the shallows around this little peak, rising out of the waves, the coral-polyps began to build what was *then* only a fringing-reef, close to the shore all round the island, except perhaps just where a stream of fresh water ran out to sea and hindered them. That is the manner in which fringing-reefs are formed, either round an island or along the mainland.

The sea-bottom sinking year by year, very slowly yet steadily, carried down the mountain, and thus the little island with its fringing-reef sank also. The polyps continued busily building up their coral-bank to the level of low-water; but as the land subsided, a channel of slowly-widening water ran between the coral and the shore. So the fringing reef was turned into a barrier-reef. And this is how barrier-reefs are believed in many cases to have been formed.

Still the centre island, with its surrounding barrier-reef, went on sinking lower and lower, very gradually, yet continuously. The little island grew smaller and smaller, but the channel of water between it and the reef grew wider and wider.

At length mere tiny peaks showed like tips in the

centre. Then they disappeared at high-tide. By-and-by they were visible only at low-tide, and soon they had sunk below even the low-tide level. The barrier-reef thus became a circular reef, enclosing a little rocky pond of salt water. The rocky pond deepened, and shells and ground-up coral, together with newly-built coral, overspread the bottom, until at length the peaks were quite buried under branching coral and white sand—and an Atoll with its lagoon was fully complete.

Whether this theory—for it is as yet only a theory —serves for the mystery of the Maldive Islands, is not quite clear. So far as the ordinary atoll and reef are concerned, it seems to be a sufficient explanation.

The thought has been suggested that, but for some such slow and long-continued sinking, the Pacific would hardly continue to this day so bare of land, with its millions upon millions of coral-polyps ever at work, not to speak of volcanic island-building in many parts.

Some atolls have been found seemingly in an earlier stage of the above history of their growth— as, for instance, with a real island rising in the centre, of peaks not yet buried.

If the ocean-bottom does thus sink, it must be at a

very slow rate, not faster than the polyps are able to build up their coral; otherwise all the islands and reefs would soon disappear beneath the waves.

From this proposed explanation it will be seen how coral-polyps may live only within a hundred and twenty feet of the surface, and yet how coral banks may reach downwards through thousands of feet.

For once upon a time, if the explanation be true, that part of the coral now so deep down, lay near the surface of the ocean. As it sank lower the coral-polyps died by thousands, and the dead coral, ground into powder by the waves, became cemented into hard limestone; while higher up the still living corals carried on the building work, only in their turn to be borne downwards, to die, and to be succeeded above by fresh generations of animals.

The speed at which coral may be formed is very uncertain, and very difficult to find out. Some islands and reefs are plainly receiving additions year by year. Others again are known to have been at a standstill for years, or even for centuries past.

If the rate of growth in one spot could be definitely settled, this would not prove that the rate there or elsewhere is now, or has been in past ages, always the same.

Varying circumstances, such as the depth or shallowness, the warmth or coolness, the roughness or smoothness of the water, also its freedom from sediment and the amount of sunlight admitted, would greatly help or hinder the advance of coral-building. Moreover that which would help one kind would hinder another, since the kinds of coral and their manner of growth differ greatly. To make allowances in any calculation for all these possibilities is not easy.

In one island a ship's anchor could be seen lying under water at a depth of seven fathoms. It had belonged to a ship, wrecked fifty years earlier. The anchor was encrusted all over with coral, yet not so thickly as to hide its shape. This was slow growth.

On the other hand, there was an islet in the Maldives having a fringe of cocoanut-trees upon it. The island was to a great extent washed away by some change in the ocean-currents, all the trees disappearing. In a few years a coral reef was built up upon the remains of the old island, entirely covering it.

Also in Madagascar certain experiments were carefully made, and proof was obtained that coral may grow, under 'favourable circumstances,' at the rate of no less than *three feet of thickness* in about *six months*. This says much!

How long the wide reaches of coral-banks and the multitudes of coral islands have taken to be built up from the ocean bottom we cannot tell. We only know that the rate of their growth lies beyond our power to determine. We only know that year by year these little creatures toil busily on, carrying out, all unconsciously to themselves, the plans of the Divine Architect.

CHAPTER XXXI.

STALACTITE.

'I have made the earth, the man and the beast that are upon the ground.'—JER. xxvii. 5.

THE Peat-formations or Peat-mosses of temperate countries ought not to be passed over without mention. Many fossil-remains have been found in them; and they offer as good an example of slow growth and gradual preparation on land as coral-building offers in the ocean.

There are in some places Peat-mosses forty or fifty feet in thickness, and about fifty miles long by two or three broad. One-tenth part of Ireland is said to be covered with peat-mosses. They are made up of half-decayed vegetable matter, piled thickly together in damp and swampy ground. Peat is in fact a kind of imperfect coal, now and then coming very near to being true coal.

The making of peat is believed to have been, as a rule, slow ; yet here, as in other matters, the speed greatly varies. The Hatfield Moss in Yorkshire was a forest eighteen hundred years ago, and it still holds specimens of tall fir-trunks. On the other hand it is recorded* that in the west of Ross-shire a considerable extent of land was, in the seventeenth century, changed in the course of only forty-eight years, from a forest into a peat-moss. This at least shows great uncertainty as to the rate of formation.

A means by which change in the Earth-crust occasionally comes about, and that suddenly, is through land-slips.

Sometimes a large mass of earth, bearing with it trees and houses, will slide down a mountain-side for a considerable distance.

Sometimes also, if a lower clay layer becomes very much softened by heavy rain or by the work of underground springs, the weight of the earth above will squeeze it out, the said earth sinking down into its place. Now and then an underground cavern, dug out by water, will suddenly collapse or close ; the ground above sinking in consequence.

Many such cases have been known, and traces of the catastrophe are often visible long afterwards.

* ' Philosophical Transactions.'

In 1806 a terrible slide happened in Switzerland on the Rossberg. A large mass from the mountain-top slid downward, avalanche-like, burying several villages and spreading itself over many square miles of country. Such a landslip as this must leave its marks through centuries following.

A few words, before the close, upon the subject of Stalactite and Stalagmite Caverns.

In the chapter upon human remains it was stated that such remains have occasionally been found in caves, together with more numerous animal-remains of different kinds.

These caves are commonly in countries where lime-stone-rock abounds. They are found in parts of England, of France, of Belgium.

The hollows—sometimes of very considerable extent, and connected one with another by long passages—were originally dug out by the action of underground streams, the Carbonic Acid in the water helping to wear away the rock. Probably after the digging-out period, a time followed during which rivers and streams flowed through the caverns, bringing thither supplies of sand or mud or other deposits, together with occasional animal and vegetable remains—fossil-plants and fossil-bones. Later still, through changes in the country—either sudden or slow, either caused by fire underground or by water above ground—these streams may have been diverted into fresh channels, and the caves may have been left almost dry.

Almost, but not quite. Water dripping gently from the roof has, in many instances, formed thereafter curious cones and cylinders and icicle-shapes of different sizes, hanging downwards in a variety of graceful forms.

Water alone would have no power to do this, but I have already spoken of abundant limestone-rocks near at hand. The dripping water, carrying a supply of lime from the said rocks, gradually drops or deposits this lime, and thus the lime-made cones and cylinders and icicle-shapes slowly grow.

The downward hanging forms are called Stalactites, and very beautiful they often are. Fine examples may be seen in the Stalactite caverns near Cheddar.

The same description of lime-formation, left by trickling water, often covers the whole cavern-floor, and it is then called Stalagmite.

The bones of animals, and more rarely of men, found in such caves, are very commonly buried in or under the Stalagmite floor. It then becomes a question of interest how long the bones have lain there.

Now, of course, if we could say precisely how

STALACTITE CAVERN.

quickly the Stalagmite was made, and could measure the exact depth of Stalagmite above any one bone, and could also be perfectly certain that the said bone had remained there quite undisturbed since the day when the animal died, we should then be able to calculate pretty closely how long ago that particular animal had lived.

But unfortunately for such calculations, we have no such steadfast foundations to build upon.

Though the thickness of stalagmite above any one bone is easily measured, we cannot be at all sure that the said bone has remained undisturbed since the animal died. It is often quite uncertain whether the animal died there at all, or whether the bone was afterwards washed into the cave. Some such caves appear to have been the regular haunts of wild beasts, who may have lived and died in them; but in most cases there is great uncertainty.

Even if we may suppose the animal to have actually died within the cave, we must still allow for the possibility of later disturbances. Heavy floods, taking place at different periods from severe rains or other causes, may have completely broken up and altered the original arrangement of bones on or in the floor.*

* 'If several floods pass at different intervals of time through a subterranean passage, the last, if it has power to drift along fragments of rock, will also tear up any alternating stalagmite

Also, the uncertainty as to the rate of Stalactite and Stalagmite growth makes such calculations unreliable. For the making of Stalactite and Stalagmite, like the making of coral, is not a thing which goes on always exactly the same, century after century, but varies in speed with changing circumstances.*

For awhile the Stalagmite formation may continue steadily; but let a wet winter come, and the dripping water increase to a flowing stream; or let a dry summer come, and even the dripping cease; or let the supply of limestone slacken; and in each case the same result follows—stalagmite ceases to form.

To examine the stalagmite in a cavern, and to note the amount of its increase between two visits, is one

---

and alluvial beds that may have been previously formed. . . . . As the same chasms may remain open throughout periods of indefinite duration, the species inhabiting a country may, in the meantime, be greatly changed, and thus the remains of animals belonging to very different epochs may become mingled together in a common tomb.'—LYELL.

* 'It is necessary to the formation of stalagmite that only so much water should be present as suffices to hold the carbonate of lime in solution. No deposit, therefore, takes place, if a stream be continuously flowing through the cavern; and even if a coating be deposited during a season of drought, this may easily be broken up again, if changes in the underground drainage of the country, or a rainy winter, cause the cavern to be again flooded.'—*Ibid.*

thing. To assert that because it has grown so much in such a time, therefore it has always grown at the same rate in the past, and therefore, again, the whole layer has taken precisely so many years or centuries in forming, is quite another thing! The first is the assertion of a proved fact. The second is the assertion of an unproved theory—and of a theory for which no proof is possible.

A good many guesses have been made as to the possible speed with which stalagmite may have been made in certain caves, where it has been found covering animal bones.

For example, it was suggested in one instance that one-twentieth of an inch in two hundred and fifty years, or about one inch in five thousand years, might be the probable rate.

Even if it could be proved, however, that the stalagmite in that particular cave had been forming at this very slow rate in late years—a doubtful matter, since some are inclined to think the growth there has long stopped altogether—still we should have no proof whatever as to the quickness or slowness of its formation in the past.

For examples of more rapid growth, as also of varying speed, are by no means rare.

Certain deal boards were left exposed to such

drippings near Durham. In the course of fifteen years, the stalactite encrusting their edges had become three-quarters of an inch thick.

A gas-pipe was left thus exposed in Poole's Hole, near Buxton. In six months one-eighth of an inch of stalactite was formed upon it. After that the growth went on more slowly, increasing to nearly one inch and a quarter by the end of eighteen years.

An iron nail was left in a forsaken lead-mine, where it caught a stalactite drip. In seventy-five years a quarter of an inch was formed upon it. Moreover, it is said that modern bottles have been found beneath a stalagmite floor, as deeply buried as mammoth-bones elsewhere.

These facts simply serve to show the great irregularity of stalagmite growth, and the uncertainty of any calculation founded upon a supposed regular rate of increase, since such a regular rate plainly does not exist.*

---

* The case of the Borness Cave in Scotland, most minutely examined by Mr. Clark, Mr. Corrie, and Mr. Hunt, may be cited as an example of this uncertainty. In the 'Final Report' on its exploration, by W. B. Clark, M.A., etc., issued June, 1878, we find as follows :—

'To the argument of antiquity, derived from the accumulation of stalagmite, I attach no great importance. Whatever value may be attached to depth of deposit over large areas, as a test of age, but little reliance can be placed on deposits over small spaces

Thus step by step, briefly, as was needful, we have followed out the manner in which, to all appearance,

of ground. The phenomena observable at numerous so-called dropping wells, should place us on our guard against any such error. Sticks at such places are coated an eighth of an inch in thickness in a few months. And, as tending to show that the deposit of stalagmite in this instance was not of slow growth, the occurrence of a cast of a piece of stick in its centre may be mentioned. Had the deposit been a slow one, the stick, not submerged in water, but exposed alternately to wet and dry, damp and cold, must have rotted long before it could become embedded. As it was, it became rapidly covered over, and by the subsequent percolation of the water through the porous stalagmite, was gradually dissolved out, leaving only a cast to mark its former situation.'

A footnote gives a somewhat contrary opinion :—

'It must here, however, be added that Mr. Hunt, after a most careful examination of the case, has come to the conclusion that the amount of stalagmite is an evidence of considerable antiquity; and that it is so, because there is not sufficient lime in the rock of the district to account for the rapid formation of the stalagmite. That this is a difficulty, I admit ; but the occurrence of stalagmite in nearly all the caves of the neighbourhood affords, I think, evidence of lime sufficient for its rapid formation.'

Again, another observer, H. J. Moule, Esq., writes to me on the subject, founding his opinion on several visits to the cave :—

'It is to be remarked that the rocks overhead and slanting from inland to the cave do not seem to have ever been greatly different to what they are now ; that the blue stone composing the mass of them is very poor in lime; and that the veins of calcareous spar, permeating them, are of no great size, and appear where visible to be as solid and undissolved as the day when they were deposited. In short, the effect on my mind was a conviction that the mode of formation of stalagmite is very imthe Crust of the Earth was fashioned by the Creator; the manner in which, through ages past, He formed it to be the Home of Man; the manner in which He still moulds and alters it, here or there, as He sees fit.

There is and must be very much that we cannot understand in the science of Geology. Nor will any really honest mind, still less any really great mind, hesitate to acknowledge the fact, and to bow low in conscious ignorance before the might of Him who alone knows all things.

We are but spelling out the broken sentences of the rock-volume, written, as it is, in a strange language, with many missing paragraphs. What marvel if we make some mistakes?

But with patience and caution we may still press on. In the great Book of Nature much may be learnt about the God of Nature. Illegible though parts of the volume may be, yet, if we read in a loving and humble spirit, we shall find some lessons to be clear;

---

perfectly understood. It looked to me as if all the lime that could be affected by the drip of water, even if dissolved out—which, as far as I could see, it is not—would hardly make up the mass of stalagmite; and, consequently, that no number of centuries or millenniums would account for the formation, on any theory that I know of.'

These differing opinions of competent observers at least help to show us how rickety is the foundation upon which age-calculations are often reared.

we shall find ourselves better acquainted than before with the boundless power of Him who 'doeth according to His will in the army of heaven, and among the inhabitants of the earth; and none can stay His Hand, or say unto Him, What doest Thou?'

THE END.

BY THE SAME AUTHOR.

Ninth Thousand, with Tinted Illustrations, cloth, 5s

SUN, MOON, & STARS.
A BOOK FOR BEGINNERS.
With a Preface
BY THE
REV. C. PRITCHARD, M.A., F.R.S., ETC.
SAVILIAN PROFESSOR OF ASTRONOMY IN THE UNIVERSITY OF OXFORD.

"Professor Pritchard warmly praises the book in a preface, and it is so nicely illustrated and agreeably written, that Miss Giberne has almost persuaded us to begin the study of astronomy on the spot.' —*Saturday Review.*

"We look at it from the unscientific point of view, and find it as worthy of praise for clearness, simplicity, and freshness of interest as the Professor finds it correct with respect to its science."—*Spectator.*

"The style of the book is well-fitted to excite attention ; it must be a dull boy or girl who will not find it attractive. The volume is one which ought to have a place in village libraries and mechanics' institutions."—*Pall Mall Gazette.*

"On glancing over the pages of this little volume, which has no pretensions further than as a first book of astronomy, we were much struck with the general astronomical accuracy pervading the work, from the first chapter to the last ; and though of necessity only the barest elementary outline of the great results of modern astronomical research is given, the authoress is evidently at home in her subject, and she has availed herself of the materials found in the most recent editions of reliable treatises on astronomy."—*Professor Dunkin in* "*The Observatory.*"

"A very charming little book."—*All the Year Round.*

"A popular treatise on one of the most fascinating of the Sciences, introduced by the hearty approval of so distinguished a critic as the Oxford Professor of Astronomy, has irresistible claims on the attention of the public."—*Manchester Examiner.*

Price 5s.

SWEETBRIAR ; OR, DOINGS IN PRIORS-THORPE MAGNA.

SEELEY AND CO., 54, FLEET STREET, LONDON.

www.ingramcontent.com/pod-product-compliance
Lightning Source LLC
Chambersburg PA
CBHW020325240426
43673CB00039B/922